Back to the Land

Back to the Land

Alliance Colony to the Ozarks in Four Generations

Ruth Weinstein

The Alliance Heritage Center
at Stockton University
2020

Published for The Alliance Heritage Center
by The South Jersey Culture & History Center Regional Press

Stockton University
101 Vera King Farris Dr.
Galloway, New Jersey, 08205

ISBN: 13: 978-1-947889-98-9

The Alliance Heritage Center at Stockton University works with community
members, academics and students to preserve and tell the story of The Alliance
Colony, the first successful Jewish farming community in the United States.

stockton.edu/alliance-heritage/
stockton.edu/sjchc/

Dedication

Once I understood what a story was, strands of my mother's family history in the old country and their emigration to the United States were well braided into my own idea of who I was and where I belonged in the world.

This memoir of my place during the years of 1943 to 1959 in the community known as Alliance (or Norma), New Jersey, is dedicated to the memory of my grandfather, John H. Levin (1883–1966), and my grandmother, Bessie Barish Levin, whom I barely knew; my parents, Sylvia Levin Weinstein (1908–1979) and Meyer "Marty" Weinstein (1902–1970); my brother, Michael L. "Mickey" Weinstein (1934–2012); to my second cousins, the Levin descendants, with whom I have renewed contact after many years; and to the efforts of my second cousin—once removed, and his wife, William and Malya Levin, to give Jewish farming in that community a rebirth.

Alliance is a creation myth for many, a Genesis in which "Let there be South Jersey" has become, generations on, "let there be America"—and it was good.

Joshua Cohen

Contents

Back to the Land

Foreword

The story of Alliance Colony is an amazing chapter in the history of American immigration. In 1882, at a time when Jews had been deprived for centuries of the right to own land in Russia and the mere thought of a successful Jewish farmer in America was unheard of, a small group of poor Jewish families escaping persecution and with no farming experience, wearing funny clothes, speaking an unintelligible language, following religious and ethnic customs unknown to the local community, arrived in a small setting in Salem County, New Jersey. They cleared the land, planted, reaped, grew in numbers, adapted, sacrificed, educated their children and overcame obstacles to persevere and survive for the sake of their future generations. They were the first Jewish farming colony to succeed in America. They succeeded with the help of their Christian neighbors, whose respect they earned. They succeeded with the help of their philanthropic cousins in New York and Philadelphia who had immigrated earlier from Western Europe and had prospered, but never forgot their obligation to their less fortunate brethren.

As the colony entered the twentieth century, the children of the pioneers, Americanized and educated, began to leave their small village. They took with them the rich and worthwhile qualities of their childhood, returning often to the village where they grew up and passing on to their children the memories and spirit of a way of life, fast vanishing. By the middle of the twentieth century, Jews again were being slaughtered in Europe. Those who remained in the colony welcomed newcomers who found refuge, comfort and opportunity where synagogues remained, Yiddish was still spoken, and a familiar spirit of Jewish community was open to them. Once again, this peaceful, rural spot on earth became a place in which to plant roots and seek the American dream.

It was of that time that Ruth Weinstein has written. As a poet, her ability to capture the subject of her prose is simply amazing. I too was there during that time. The aroma of the hotdogs sizzling at Norma beach, the sounds of the juke box playing in the pavilion, the laughter and shouting of the people, the splash of the water at Dopey's, and all of the other sounds, smells and sights are brought to life. The goodness, fine qualities, wholesomeness, purity and kindness that imbued that community and seemed to linger over it are conveyed to the reader in prose, so well-crafted as to somehow return us to that simpler time.

My personal thanks to Ruth for having taken me back with her. Her memoir is a tribute to those who settled a small part of the world in which new life blossomed with a lingering sweetness. She describes a time and place that we were fortunate to experience, and that we fondly and wistfully remember.

Jay H. Greenblatt

President,
Alliance Colony Foundation

Chapter 1

Why on Earth Write a Memoir?
An Introduction

Invariably, when I meet people in Arkansas and they hear my mid-Atlantic accent, they ask the same questions: "What brought you to Arkansas?" and "Where did you grow up?" To the first question, I answer, "cheap land" and to the second "northeast, row-house Philadelphia, with summers in rural, southern New Jersey." For longer than I can say, I have wanted to write the story of how I got from my roots to the Arkansas Ozarks by shining a light on the brief span of my summers in Norma, New Jersey. Now close to offering it to the world, I wonder who would want to read such a memoir. I am a relic of history: an old, white, cis-gendered, straight, Jewish woman, who while writing this in 2018 and 2019 has been married to the same man for nearly forty-three years. I have suffered neither alcoholism nor drug addiction, mental illness nor physical disability, sexual nor physical abuse. I am relatively healthy and fit, considered intelligent, but have no stellar achievements on my resume, and am not charismatic. I am altogether too ordinary for my life to garner interest when memoirs abound of people who have overcome almost insurmountable odds, people with major accomplishments, and people who reveal horrifically painful pasts from which they soar to great heights.

My maternal ancestry, however, is part of a fascinating piece of American history: the revelation of the link that explains the atypical American, back-to-the-land lifestyle I have embraced for over forty years—more than half my life. I am a descendant of one of forty-three families who share a certain distinction. In 1882, the original settlers of The Alliance Colony—our ancestors, our dear pioneers—all emigrated to escape the pogroms of Imperial Russia waged by Tsar Alexander III against the beleaguered Jews of his empire. Their arduous journey took them, finally, to the sandy soil of southern New Jersey with the purpose of becoming farmers. They would not only change their location but nearly every aspect of their lives since, with few exceptions, Jews had not been able to live as farmers in Russia for centuries. First settled in May of 1882, Alliance (and local, offshoot communities) existed for nearly a century, sending descendants out into the world to be successful in every conceivable field of endeavor.

Among the first generation of children born in the colony are several whose achievements were well-known. Moses Bayuk's sons established the Phillies Brand Cigar Company in Philadelphia; Joseph B. Perskie became an Associate Justice of the New Jersey Supreme Court from 1933 to 1947; and the Seldes brothers were both writers of significance. Gilbert, the older, was a writer and social critic, while his brother, George, was an investigative reporter, foreign correspondent and editor. Those who are a part of the Alliance Community, a loosely connected tribe of sorts, go way beyond the descendants of the original forty-three families, as other seekers of freedom in America swelled the numbers of the tiny communities of Alliance, Brotmanville and Norma, New Jersey, known collectively as The Alliance Colony.

Alone my treasured family history would not be significant; many of the descendants of my generation are highly accomplished, and like their ancestors have produced a handful of American luminaries. I, however, took the literal message and history of my ancestors' migration to heart. After college graduation, I became an English teacher in the Philadelphia school system for about eight years but always struggled to find fulfillment and happiness. Memories of my childhood and teen summers spent in the bosom of this remarkable and nurturing community left a mark on me—a brand in my consciousness—and tugged at an unrequited longing. In the late 1960s and early 1970s, the back-to-the land movement, with strong parallels to my ancestors' lives, was stirring across America. I travelled to California and Montana hoping to find people with whom to buy land. Nothing clicked until, in 1974, I met Joe McShane, an Irish-Catholic boy from a Philadelphia, working-class background like my own parents' socio-economic status. Joe was imbued with the same pride in his mother's farming roots in western Ireland as I was by my mother's family's origins. We were destined to live a life together with an awareness of his family's struggles against British colonialism in County Mayo, Ireland, and my family's goals of becoming "free farmers on their own soil" in Pittsgrove Township, Salem County, New Jersey. Together we have made a modest, a long and good—though very rarely easy—life on forty hardscrabble acres in Searcy County, Arkansas.

Tomatillo Ridge Farm, 2019

෨

And So I Write

In one way or another, I have been writing about being a summer-kid granddaughter of that community all my life. As a school kid, I must have written numerous enthusiastic compositions on the subject of "How I spent my summer vacation." There is a sweet, hand-written poem I wrote and dedicated to my mother and the memory of my father in 1971, five years after my grandfather's death and a year after my father's, extolling childhood summers in the country at my grandfather's bungalow. In the 1980s, I filled notebook pages, also by hand, in an attempt at an autobiographical, coming-of-age novel which I glance at now and then. I cannot yet bring myself to compost the pages nor use them as fire starters in the woodstoves that heat our house. I became serious, if sporadic, about writing this memoir after several years in the early twenty-teens during which I wrote a lot of poetry, a Jane Austen-kind of poetry, daily meditations upon the themes of conjugal domesticity, gardening, nature, and community—themes in a way engendered by my early New Jersey experiences.

In 2015, my longing to travel back, one more time, to the land of my youthful summers became so palpable that the slightly uphill path to my garden in the long shadows of summer's late afternoons could fool me into feeling that I was sixteen and walking the much longer, but flat, path through the woods to the South Jersey swimming hole of my teenage summers. I despaired about ever getting there again since, even if still alive and *compos mentis* for the 150th anniversary of the founding of the colony, I would be ninety-one and probably unwilling to subject my body to the ordeal of long-distance travel. I could only free myself from the obsession by committing to write about it, my efforts initially neither cohesive nor graceful. Further research and disciplined writing were the only methods by which I could give form to the partial memories that occupied me on a cellular level, that were evoked by all my senses, and that I had to bring alive once more.

The fact that there are no immediate relatives left in my life whose experiences overlapped with mine in those times and in that place matters deeply. My beloved brother, though seven years older than I, shared many of the same experiences, but he died in the spring of 2012, having spent the last seven years of his life as a tetraplegic after a bicycling accident paralyzed his previously very fit and healthy body. Mickey had always been my protector and best friend. I had an older cousin Bruce who shared some of our childhood summer influences; he is long deceased. I am not in touch with my two remaining first cousins—a pair of brothers, one my age and the other eleven years younger. My recent attempt at restoring communication with my peer through a brief email exchange ended leaving us both unsatisfied; memories of childhood, for him, seem to be a provocation devoid of pleasure. I apologize for any pain I may have caused him when I asked him to recreate parts of our past with me. May he live his present moments in peace.

Although my younger cousin and I were once close, his childhood and mine were not contemporaneous. His assistance in building the house in which my husband and I have lived for nearly forty-two years—after a fire destroyed our first Arkansas dwelling—has made his existence in my life a lasting, material one, despite our lack of contact now. Like me, he has made a life more akin to that of our ancestors than any of the first, or even our second cousins. He, like our great grandfather, Israel Hersh Levin, became a highly skilled and successful furniture maker/finish carpenter. I see us both in an atavistic light, as sports of nature. *Webster's New Collegiate Dictionary* defines this seldom used meaning

17

of "sport" as "an individual exhibiting a sudden deviation from type beyond the normal limits of individual variation, usually as a result of mutations, esp. of somatic type." I use the term "sport" more figuratively than literally, not so much a somatic sport, not a physical deviation, but rather a psychic or cultural one. While so many of my cousins have had successful careers in various fields—the law, academia, medicine, social work, business, education, the arts, and more—I'm the sole one of my generation who has turned so whole-heartedly to the soil. I see myself as one of the wild ones, needing, in order to survive, a close, daily connection to the land, more than anything else beyond the basic, physical necessities of life. I cannot thrive on concrete and macadam.

In my old age, I have found a comfortable balance between my need for social companionship and solitude. My solitary aerobic walks most mornings and my occasional, solo yoga sessions in the afternoon are satisfying to me, more so than exercising with friends or driving too far to take generic yoga classes. It helps that I am self-motivated. Because I often read books many readers have no interest in reading, I have a somewhat anti-social attitude about book clubs. I have a wide variety of friends, many young enough to be my children and grandchildren, a few older. I am deeply touched that several friends find me to be a comfort and an inspiration. Although in the past I have experienced times of bitter loneliness, that has not been a problem for years. However, one of the few curses of old age that I feel is a kind of desolation that goes along with being the sole inheritor or caretaker of a family's story. That my first cousins do not wish to share these memories leaves me feeling cutoff and melancholic that I alone treasure those reminiscences, but fortunately, many friends, acquaintances, and even strangers find my recollections of this past fascinating and encourage me to share them through the vignettes I craft of that time in my life.

One of my closest friends describes me as living an "art-driven life." I am grateful that she knows me so well. I include in this all-encompassing drive the need to write; to make woven, quilted and sewn things of beauty and comfort; to paint; and to garden and cook food that nurtures the body, as art nurtures the mind and heart. Mostly though, gardening drives my life with its seasons and climate, weather and soil, miracle of sprouting seeds and beneficence of harvest. So often while on my knees weeding or picking, poems come to me among the peas and beans. For nearly forty-five years, my husband and I have grown organic gardens, and if I exit this life with a shred of luck and grace, I will die in my garden. Gardening has holistic, medicinal benefits more people need in order to become whole in this world.

Another factor that encouraged me in this process of writing a memoir was meeting Mike Jennings, a retired journalist and Vietnam Vet, who was in residence at the Writers' Colony at Dairy Hollow in Eureka Springs, Arkansas, in the autumn of 2016 to work on his war novel. Not his protagonist, but one of his major characters, a New York Jewish lawyer, spun the tale of his Polish-Jewish ancestors and the Polish countess who saved them from a peasants' pogrom against Jews in Odessa in 1905. Mike read the harrowing account to the group gathered at the monthly "Poetluck" potluck. Then before I read a few poems, I told the group, who mostly knew me as a poet, that I had begun a memoir/family history and spoke briefly about my Alliance roots. An unmet need of this writer's novel and the bounty of my family history became the basis for our email epistolary friendship as I helped him give his character Rabby a home in a fictional Alliance Colony to back up the character's true origin story—entirely different from the one Rabby spellbindingly told his fellow soldiers on a rooftop in Vietnam. That kind of sharing with another writer feeds the soul and fuels the incentive to keep writing just as much as how I live my life does.

❧

Going Home Again

In mid-August of 2018, I returned to that small community of my family's roots for a celebratory picnic honoring the 136th anniversary of the Colony's founding. In my lifetime, there have been three such celebrations. Traditionally, *yovals*, or jubilees, are held every fifty years, but a little bending of tradition in scheduling brought great joy to many people. A grand celebration for the Centennial, complete with a beautiful commemorative poster—a copy of which a cousin had sent to me and which is still a treasured possession—drew over a thousand people in 1982. My brother, his wife and various cousins attended. I would have loved to attend, but my husband and I had other plans.

For ten weeks that summer, we traveled around Ireland while a young couple and their baby stayed at our little Ozark homestead, milking our goats, tending our chickens and pets and harvesting what remained of the spring garden we had planted. It was our honeymoon, delayed by six years and not in the least luxurious, since we hitchhiked often and camped part of the time. We stayed with my husband's family, as well as with strangers who gave us rides, meals and warm hospitality. We stayed in modest bed and breakfast establishments and with a family we had corresponded with because, like them, I was a handweaver. Years later, in one of those weird, surrealistic, small-world, cosmic crashes, my husband met their daughter again at a craft show at Berea College in Kentucky where she and her brother, who was not at the show, were studying.

The 125th reunion occurred when the officers of the Alliance Colony Foundation and other organizers decided that they "could not afford to wait that long (another twenty-five years) as memories fade and familiar faces disappear."[1] I did not attend that reunion; at that time, Alliance did not so prominently fill my radar screen. In 2007, my brother was two and a half years into his life as a tetraplegic, and family focus meant visits to him. Fortunately, the same urgent sentiment of the aging descendants to gather again led to another picnic on the lawn of the old Alliance Cemetery a mere eleven years later. I received the invitation to the 136th Alliance Colony anniversary picnic because I had begun intensive research on Alliance in about 2013, when the desire to write a memoir was inchoate, through an occasional email correspondence with the president of the board of the Alliance Colony Foundation, Jay Greenblatt.

So yes, I was there in mid-August of 2018 during the height of summer gardening, on a five-day whirlwind trip including the hundred mile drive each way from home to Little Rock, flights to and from Little Rock and Washington, D.C., with brief respites at the suburban Maryland home of my sister-in-law. Sandwiched between the fast-paced flights and family visit, I rented a car—a first in my life—and drove from Rockville, Maryland, onto I-495, for what, in ideal traffic conditions, would be less than a three-hour drive to a tiny vestige of a once thriving community in southern New Jersey. Such ideal traffic conditions on I-495, I-95 and the Delaware Memorial Bridge are figments of the imagination, and when nearly four hours later I approached familiar territory, the sky darkened and released rain in obliterating, drenching sheets. I was there for less than forty-eight hours and drove again from the ancestral homeland very early Monday morning in equally torrential rains to catch a 7 p.m. flight to Little Rock. By Tuesday afternoon I was back in Arkansas, exhausted but almost manic with excitement. I was on fire about the picnic, about reuniting with second cousins I had not seen in sixty years, about meeting people whose family names have been familiar to me all my life—whether I had known the individuals I was greeting or not—and younger, but kindred spirits, who were also descendants. There were documentary filmmakers, genealogists, artists, musicians, farmers, professionals of all stripes. While enjoying the reunion in Alliance, New Jersey, I was amazed when I met a Philadelphia gentleman roughly my age who had only recently learned of his family's history through Ancestry.com. Like most of the descendants among the 450 attendees—especially the older ones—and many who could not travel for the event, I had imbibed our history along with mothers' milk.

Often in old age we return to the memories of youth, but we need not do so in a miasma of dementia. In his 2009 volume *The Shadow of Sirius*, W. S. Merwin, the great American poet, set forth a collection of exquisitely rendered memories stripped of sentimentality. We are treated to his recollections of conversations, events, observances of light and other sensations, all striking to the child who stored them carefully. Now as an old man he plays his concertina of poetry, and with each push and pull of the bellows, out comes a perfect note from the rich storehouse of his past. He inspires me to reach back to the treasured past of my childhood, basking in the warm glow of the Alliance Colony's dwindling days and to pull out illuminated moments and experiences to share.

Through the process of writing a memoir, one learns much about who one really is. The result cannot be laid out for the reader like a tablecloth upon which a complete meal, from appetizer to dessert, is served in an orderly manner. Sometimes it is necessary to interrupt the story of the Russian Jews in total or my family's history in specific with the sudden flashes of self-realization. I cannot offer you a tidy, chronological account of a whole tribe of people, a large extended family of multiple generations, and an old woman looking back at her first eighteen years to understand how the stories contained within this generational history have shaped her life. Friends who know me primarily as a poet, along with my own writing practices, have encouraged me to include poems in the memoir. As I read parts to these friends or email sections for them to read, they say things like "oh, the tomato poem should go here" or "it feels like a poem needs to pop up in this section," and I have decided that they are right.

Readers, if you are game to spend the time, come wander on a path with me, a path through a long life, meandering in a forest glade where sunlight illuminates spots on the ground or clumps of leaves on ordinary trees. We may pick up and turn over a stone and hold it lovingly in our hands or toss it into the brush as unimportant after all. Though it seems like any path through a copse of trees, we will find and explore treasures of memory, fragile and intricate as a bird's nest, glittering as flakes of mica, shy as a woodland animal, bold as a thunderstorm. Come with me as I dance back and forth between childhood and my family's history, the long history of my people and an art- or gardening-driven day in the present. My wish is that my memoir gives you pleasure and invites you to walk the paths of your own lives with new eyes, sifting through the memories to find your own treasures and share the stories they tell.

Back to the Land

Chapter 2

The History of the Alliance Colony

Russian Jewish Life to the 1880s and the Pogroms

My intention in summarizing the history of the Russian Jews from the time and locations of their persecution in the old country to their settlement in the new world is not to add yet another scholarly work to the canon of academic material on the subject. I could not possibly add anything of significance to this body of work. Historians of the Jewish-American and New Jersey experiences and descendants of the Alliance settlers seem satisfied with what has been collected and written, and others who are more qualified as historians write additions as they see fit. Rather, I wish to provide an adequate and varied survey to arouse the interest of people who know nothing—or little—of the effort to turn the victims of anti-Semitism in the Russian Empire of the late nineteenth and early twentieth centuries into farmers in the Americas. Also, this history provides a necessary background to my family's history, as well as my personal journey, and will, I hope, whet readers' appetites to learn more.

Since the 1600s, waves of pogroms against the Jewish people of the Russian Pale were waged by Cossacks at various intervals, but the 1861 Edict of Enlightenment, which partially freed the Russian peasantry from serfdom, also relaxed some of the long-exercised and highly restrictive laws against the Empire's Jews. It liberated them from the rules of residence which had confined them to city ghettos for centuries and from the laws which excluded them from secular and university education and unrestricted employment. For nearly twenty years, Russian Jews enjoyed a relatively improved life, as wan rays emanating from the Age of Enlightenment of Western Europe began to penetrate the benighted perceptions of Holy Mother Russia held by clerics and the Russian aristocracy. Under Tsar Alexander II, more Jews in Russia had greater access to university educations as well as to secure livelihoods as government employed artisans, minor civil servants and professionals.[1]

In 1881, Alexander II was assassinated in a bombing perpetrated by a group of young revolutionaries desiring to make Russia a modern, westernized European country. Alexander III, the murdered Tsar's son, did not share his father's more enlightened views and ramped up with renewed vigor and an iron fist the dreaded pogroms against the Jews. Destruction of property, murder, rape, military conscription and the threat of banishment to Siberia ensued. Furthermore, under the policy of juvenile conscription, the Tsar drafted into the Russian military young Jewish men, who were hardly out of childhood, with the intention of converting them to Christianity, if they survived being cannon fodder. To sow even more internal strife in the Jewish community, the government required that its leaders, the *kahal*, choose which young men were to be conscripted.[2] It is not surprising that with brutal enforcement of such restrictive laws, these young men and their families were unwilling to serve what amounted to life sentences in the military. Whenever possible, Jewish families would pay as generous sums of money as they could afford to put their own young sons' names on the death certificates and graves of young Christian men. In this way, some were able to escape conscription. But sometimes the ruse would be discovered and met with severe punishment.[3] In desperation to keep sons out of the military, some parents cut the tendons of their sons' index and middle fingers on their right hands so that the two "trigger fingers" could not function.[4] Only the protection of their more compassionate and enlightened Christian neighbors helped the Russian Jews to survive the pogroms.

The new Tsar's violent policies against the long-victimized Jews of western and southern Russia were whole-heartedly supported by Russian clerics and aristocracy who feared inroads on their privilege. The effects of industrialization that were taking hold in western Europe—and that would forever change life for the whole world—contributed to the worsening of conditions for the Jews rather than to the improvement. They received no benefits from the weaker form of industrialization that was just beginning in the Russian Empire, and no access to better employment. In fact, encroaching industrialization eliminated some of the traditional jobs available to lower and middle-class Jews, such as wagon driver, inn keeper, artisan, miller and estate agent to the nobility.[5] The opposition to reforms that would make significant changes in Russia allowed violent hatred directed against the Jewish people to flare, which in turn led to a massive exodus from the Russian Pale of Settlement, the greatest migration of Jews since the Diaspora. The Pale is that area to which Jewish habitation was restricted. In the 1880s, nearly eighty percent of the world's Jews (almost six million) were confined to sections of frontier borderlands of Russia, Poland, and Austro-Hungary, now known as the Russian/Polish Pale, although most of it is in modern-day Ukraine. (While I refer to this area as the Russian Empire throughout this book, I am aware that Russia invaded Ukraine then as it has in current times.) An estimated 3.2 million Jews left that vast area between 1880 and 1920, with many ending up in the United States.[6]

◇

Responses to Pogroms

After the reinstatement of the pogroms, a new back-to-the land movement sprang up in the shtetls of the Pale. The young leaders of Am Olam (the Eternal People or the Eternal People of the Eternal Land) believed in "productivization" and "normalization" of Jews through agricultural labor in order to demonstrate to the world that the Jewish people could be an asset. Am Olam had a brief but inspired, and inspiring, existence. Perez Smolenskin's Hebrew essay "Am Olam" ignited three young utopian idealists—Nicholas Aleinikoff, Michael Bakal and Moses Herder—to develop policies for settling Jews on America land in socialist communes. Am Olam groups popped up in Kiev, Odessa, Ylizavetgrad and elsewhere, studying and planning for their grand experiments. If they could not be assimilated into Mother Russia to work the land there, they would build Zion elsewhere.[7]

The first Am Olam group to emigrate in 1881 consisted of thirty-two idealistic young families, most from Ylizavetgrad.[8] Their destination was one thousand acres in Sicily Island, Louisiana, which was flooded by the Mississippi River the next year. The remote location with its undeveloped, infertile land; plagues of disease-causing mosquitoes and snakes; the oppressive, tropical heat and uninhabitable buildings; and a host of other obstacles, including financial difficulties, ended the endeavor.[9] U.S. highway 425 from Winnsboro, Louisiana, to Natchez, Mississippi, passes through Sicily Island. The 2010 Census listed 526 inhabitants, nearly half of whom lived below the poverty line, more than half being African American. According to the road sign, we were within twenty miles of Sicily Island in 2016 on our way to the Mississippi Gulf town of Ocean Springs to visit the Walter Anderson Museum of Art. I had just read about the ill-fated Am Olam settlement and pondered that there could not have been much to support the thirty-two Jewish families who emigrated from Russia to Louisiana then, nor much to support a shrinking population in the second decade of the twenty-first century.

After the Sicily Island colony disbanded, some stalwart survivors tried again in Cremieux and Bethlehem Yehuda, South Dakota. These colonies lasted two or three years before collapsing in financial ruin. Beer Sheba, another attempt, was established in Kansas but failed in the same way. All that remains are a few lines in Wikipedia, other on-line sources, and scholarly Jewish-American and state history tomes. The poet in me likes to think that sometimes in the near delirium-inducing temperatures of summer, brief ectoplasmic manifestations of the people with foreign looking visages and clothing and strange sounding names and speech, who inhabited the colonies for a blink of time, can be seen in the heat mirages hovering over the wheat or soybean fields in South Dakota and Kansas or the rice and cotton fields in Louisiana.

Most successful of the Am Olam-inspired societies was New Odessa near Portland, Oregon. It had a stronger ideological bent and for most of its brief span, a charismatic leader, the non-Jewish William Frey, who preached the Positivist doctrine of August Comte—the "religion of humanity." Frey's philosophy aside, he had had considerable agricultural experience, unlike the emigres who had been city dwellers. Nearly seventy settlers, mostly young men, including the socialist Pavel Kaplan, arrived in summer of 1882 and began to clear the land. The following year they adopted a constitution which made two significant declarations: it prohibited all external labor and commerce by the colonists and guaranteed equal rights for women and men. The colonists grew crops and sold timber to the railroad.

Soon enough challenges emerged. The labors of a land-based livelihood were too much for many of the settlers; the young men competed for the interest of a much smaller female population; and Frey's ideologies that veered so far from Judaism began to feel uncomfortable for some of the settlers. Frey and his wife left in 1885, and soon after that a fire destroyed the main building. In 1887, bankruptcy and foreclosure saw the end of yet another ephemeral, Jewish agricultural-communal experiment. Oregonian historians view New Odessa as significant, despite its brief existence, because it was one of very few utopian communes of that era in Oregon.[10] Our Levin family history states that the option of going to Oregon existed for the group of immigrants who arrived in New York, but they chose instead to go to New Jersey.

The oddest, and to me, the most interesting failed settlement existed for no longer than a year near the city of Newport in Jackson County, Arkansas.[11] Newport is 110 miles east of where my husband and I live in the Boston Mountains of the Ozark Plateau, but northeast Arkansas, unlike the steep terrain of the Ozark region, is flat and swampy.

> During the spring of 1883, a group of about 150 people left New York for Arkansas after receiving a tempting offer to settle on a tract of land and supply staves to a lumber company. Some of them were survivors of Sicily Island who were ready to try again. At first, the thickly forested site looked promising to the colonists, who wrote back to New York of the Eden they had discovered. Although the connection of the first group of settlers to the Am Olam is unclear, their letter was received by one of the Odessa Am Olam groups waiting in New York for just such an opportunity. A group of about thirty Am Olam members, including three families, quickly moved to Arkansas and, with their own money, purchased land near that of the first group.[12]

Kate, the seven-year-old daughter of Moses and Rachel Herder, wrote an unpublished memoir of the difficult life in this unidentified tract of Arkansas land. Moses Herder, an intellectual and idealist who strongly believed in the dignity of working with one's hands, was one of the two founders of Am Olam. Upon the failure of the Arkansas colony, the Herder family was joined by a man named Rosenthal in a journey to join the Carmel, New Jersey, colony, a "satellite" of Alliance.[13] Carmel has always been in my consciousness as an adjunct to the Alliance experience. My family talked about it and the people they knew, and I had friends from there in my youth. In the historic context, Carmel had the reputation of being more socialistic, philosophical, politically radical, belonging to the intelligentsia. Eugene V. Debs and Emma Goldman gave lectures there.[14]

Less than one hundred years later, my husband and I reversed their journey as we traveled from the farm where we were living northeast of Philadelphia and made a successful back-to-the-land life in the Arkansas Ozarks. The conveniences and pleasures of modern life in 1976, the knowledge that there were options we had freely chosen to leave behind, and a strong nucleus of kindred spirits in our new community made our experience a lot easier than that of the Arkansas Am Olamers. However, life for homesteaders in the 1970s, who were eschewing the suburban American dream, was not the proverbial bowl of cherries. I deeply understand the hardships the Am Olamers endured from my own personal experience. The effort of long days of back-breaking work to clear land for an orchard with the meager reward of a bath that consisted of a bucket of hot water in the middle of the kitchen and wash tub to stand in, or a trip to the river in the early summers (from 1976 to about 1982), was not that much better than the same work and personal hygiene resources a scant hundred years earlier.

Kate Herder's youthful account of their life is illuminating. "We lived in that house [a group dwelling] for some time

31

and we were not the only family neither. There must have been about four more besides ours." Herder reveals that the group purchased their basic needs together:

> . . . all the families scraped together all the money that they had. They coaxed the owner of the house to let them have the horse and wagon only to the store and back because they were going to buy a big supply of the most necessary things they needed.[15]

Young Kate describes participation by the entire family in the labor that sustained them.

> Father would go around with an ax and chop around the twigs and the bushes around the big trees, and we children would pick all the twigs together and make a big pile of them. And for this work that all the family did from the oldest child to the youngest including Father and Mother, we got instead of money from the man that owned ever so much land, cows, pigs, and horses, yellow corn meal, matches and the water that drips off the cheese.[16]

Herder's narrative includes a makeshift Passover meal with coffee substituting for wine. As with South Dakotan and Kansan settlers who had to accept hogs for payment, the consumption of pork had to be an enormous adjustment for the Jews living in such isolation and desperate conditions, even those who did so willing. They had to survive, and they were idealistic intellectuals rather than pious Jews. The primary reason for leaving this place—other than isolation, near-starvation, and brutally long hours of back-breaking work—was the location; the land was a snake-infested, malarial swamp where temperatures rose to 105 and 108 degrees Fahrenheit, something the health of the Russian immigrants just could not endure. Ninety percent of the Arkansas settlers came down with yellow fever or malaria, and eighteen to twenty died.[17]

That they survived these conditions and lived is notable. What the native Arkansans thought of these people who briefly lived among them we will never know. When they finally made it to Carmel, New Jersey, their relief must have been palpable. In my youth, southern New Jersians sprayed DDT to control mosquitoes, mosquitoes quite worthy of hyperbole and complaint, but compared to eastern Arkansas, South Jersey's sandy soil farmland, scrub oak and pine woods, and tea-brown waters were practically Acadian. There they had opportunity for a more fruitful experiment. And there they found others with a common background in Carmel, Rosenhayn, and the mother community of Alliance (Alliance, Norma and Brotmanville). Working in a community can make the most difficult job lighter or drive people to part in rancor. As the story of Alliance shows, both cooperation and community as well as irreconcilable differences among neighbors became part of the New Jersey settlements.

Back to the Land

Chapter 3

The Alliance Colony Benefactors

While Am Olam no doubt influenced some of the forty-three families who settled Alliance, several factors explain why this group did not, as a whole, exhibit an undying commitment to the idealistic communalism and socialism of the New Odessa commune and other groups. The families constituted a significantly larger group that came from several different areas in the Pale, and they did not all subscribe to the philosophy of Am Olam. They were located near Philadelphia and New York, from where their patrons were able to keep close contact and exercise some control over them. These urban centers also provided ready markets for their produce and supplied access to winter or auxiliary employment. Soon enough, the circumstances of life in the new land caused socialism to yield to capitalism. The fact that the Alliance settlers came from various cities and backgrounds and that most did not know each other until they became interdependent neighbors must have influenced the social and political beliefs of the day; perhaps it provided more fertile ground for competition, the hallmark of capitalism.

Ellen Eisenberg identifies at least nine of the original forty-three settler families as having strong ties to Am Olam prior to emigration, with several others developing ties in the early days of settlement. Among them are the Kiev emigres: the Levins and Levinsons, related through the marriage of my generation's great, great grandmother Leah and her second husband Berel Levinson; the Mennies and Strausniks; and other families from elsewhere in the Pale.[1] Eisenberg claims that all of these Am Olam settlers remained in the colony through 1900, but my grandparents were there at least through 1913 and possibly later, while my Great Uncle Bill and Aunt Lena remained on their farm for their entire lives. I do not recall hearing the term Am Olam in my youth, but this may be a function of a young girl's self-absorption. While I cannot remember that our family used the term, the ethos of the Levin family's connection to the land was always solid, and front and center in my consciousness.

ᕽ

Alliance Israelite Universelle

Although the wealthier, more educated and cultured German Jews of Hamburg were not welcoming to the Ostjuden (Jews from Eastern Europe), the French Jews, at least, came to the aid of the emigres with financial backing. A bias against the less educated, more superstitious, more orthodox, and poorer Jews of Eastern Europe had always existed. The colonists found support for their emigration and settlement from two primary philanthropic sources: Alliance Israelite Universelle of Paris (AIU), for which they named their colony, and the Hebrew Emigrant Aid Societies (HEAS) of New York and Philadelphia. These benefactors made it clear that they were not providing charity, that they would only support those who would not be burdens upon the philanthropic organizations, and that they had their parameters. These stringencies were to be even more enforced by the American benefactors whose money came from industrial and commercial profits.

AIU, the first modern international Jewish organization, was founded in Paris in 1860 by emancipated Jews who had prospered and assimilated in their own countries. They vowed to work everywhere they found Jews suffering from antisemitism and discrimination, to free them, to work for their moral progress through effective assistance, and to publish printed works calculated to promote Jews. The organization still exists today.[2]

∾

Hebrew Emigrant Aid Society

HEAS, on the other hand, was the least generous of the early philanthropic organizations and believed in limiting immigration of the Eastern European Jews. When it disbanded in March of 1883, its work was taken over by United Hebrew Charities. By this time, HEAS in America and the AIU in Paris were moving away from resettling the eastern European Jews in rural colonies. However, Michael Heilprin, a Polish-born Jew, scholar, author, and philanthropist, who had emigrated to the United States in 1856, still believed in agriculture as a solution to the problems of the poor Jewish immigrants. He was an active participant in HEAS during its existence and, after its dissolution, continued his ardent pleas to wealthy American Jews for generous support of the settlement of Jewish agricultural communities.[3]

Two incidental facts about HEAS are of interest to me and seem appropriate here. The high-handedness of the German-Jewish directors of HEAS and their poor treatment of the immigrants prompted Emma Lazarus, originally an enthusiastic volunteer of the organization, to become one of its strongest critics.[4]

Through HEAS, American benefactors in New York and Philadelphia, for whom many of the roads in Alliance/

Brotmanville/Norma were named, made their financial contributions. Henry Avenue, where my great grandparents lived, and Gershal Avenue, where my grandfather lived during my childhood and youth, were named after the philanthropists who aided the immigrants.

It is important to note that today, in the late autumn of 2018, the Hebrew Immigrant Aid Society, known as HIAS, is not the same as the Hebrew Emigrant Aid Society, or HEAS. Although HIAS does similar work to that done by HEAS from 1881 to 1883, HIAS is the progressive Jewish aid organization identified as one motivating factor for the senseless and horrific murder of eleven Jewish worshippers on Shabbat at the Tree of Life Synagogue in Pittsburgh, Pennsylvania, on October 13, 2018. The white supremacist, anti-Semitic, alleged shooter railed against HIAS for its work in aiding the caravan of Central American refugees walking towards the U.S.-Mexican border. Prior to the 2018 mid-term elections, the 45th president of the United States, Donald Trump, called the Guatemalans, Hondurans and Salvadorans seeking physical safety and a better life in the United States "an invasion." Established in 1881 to aid the waves of immigrants from the pogroms in Imperial Russia and evolving to help refugees from World War II, their work since 1975 has focused on providing welcome, safety and freedom for refugees and immigrants of all races, ethnicities, religions, sexual identities/preferences and regions in the world.

∾

The Benefactor Who Wasn't

The name of Baron Maurice de Hirsch was a part of the oral history of my mother's family that I have heard all my life and later read about in greater depth and more accurate detail. I do not know how his name became so mythically conflated with my family and the Alliance Colony because the connection between the famous man and America's first successful Jewish farming community is tenuous at best. As a child, I naively felt that there was something grand about our family's connections with this larger than life philanthropist. Although my great, great grandmother Leah Levinson and her daughter and son-in-law, Esther and Hersh Levin (my great grandparents), were completely unknown as individuals to this wealthy and compassionate man, he was moved by the plight of Jews like them. To the child I was those many years ago, that was a special, albeit erroneous, link. Observant Jews, and even non-religious ones who acknowledge the culture of mourning among Jews, have a saying to convey respect and condolences when someone virtuous dies. One workable English translation is "May her/his name be a blessing." Any utterance of the Baron's name that reached my ears in childhood always inspired a sense of awe. It always seemed that my family were implying that his name was a blessing to us particularly. While I deem it vital that the historical facts of this book be correct, it remains a memoir, not a history text. Because of the inaccurate but powerful mythological significance the Baron had in my family, I include a brief summary of his life and charitable work. Most likely, the story of how our family benefitted from the Baron's philanthropy was passed along through my grandmother's family, who probably came to the Alliance Colony in the late 1890s when de Hirsch took a more active and liberal role in bestowing help.

He was born in 1831 into a family of Jewish court bankers, his grandfather being the first Jewish landowner in Bavaria. His mother, Karoline Wertheimer, made sure that he had excellent instruction in Hebrew and religion. He married Clara Bischoffsheim, daughter of a Belgian Jewish banking family, augmenting his fortune considerably. However, he was enormously successful as a businessman and investor on his own merit, with interests in sugar and copper in North and South America. He invested—to the horror of some of his peers—in the Oriental Railway, connecting Vienna and Istanbul, and was greatly rewarded with the triumph of his audacious investment. His young wife had already established herself as an active benefactor, and together their contributions to Jewish philanthropy were enormous. It is considered impossible to assign anything like an accurate total to their financial gifts.

They established hospitals and schools in the Middle East and Galicia where Jews lived in poverty and ignorance. When the renewed pogroms began in the Pale in 1880, de Hirsch's attention was drawn to the Jews of Russia, although his modest donation to the Alliance Israelite Universelle made him an indirect donor. At that time the Jews of Eastern Europe, particularly those in the Russian dominated Ukraine, were despised and discriminated against by prescriptive laws. They lived in deplorable conditions. When his only child Lucien, a young man of great promise, died at thirty, the baron replied to an offering of sympathy by saying, "My son I have lost, but not my heir. Humanity is my heir." He could have spent his multiplying millions on selfish pleasures, but instead dedicated his prodigious wealth to improving the lives of his persecuted co-religionists. He helped to fund various institutions that sent Jews to agricultural colonies from Argentina to Canada, from Oregon to New York, to places with odd Hebrew names similar to the Beer-Sheba, Kansas, and Bethlehem-Yehuda, South Dakota, settlements which he did not likely fund. The Baron's vision for them supported the radical philosophy of Am Olam, of turning Jews into farmers, specifically in the western hemispheres of North and South America. It predated the geo-political-religious movement of Zionism and the idea of Eretz Israel—a modern homeland for Jews in Palestine and parts of the Trans-Jordan.

In fact, de Hirsch's generosity did not directly fund the Alliance Colony. Rather, in 1881, a trust fund of £493,000 funneled through the Alliance Israelite Universelle helped enable the settlement of the Alliance Colony. In 1891, the Baron de Hirsch Foundation was established and continued funding the exodus of Russian Jews into the 1920s with an emphasis on emigration to Argentina and Brazil. In the late 1890s he made a direct but apparently modest donation to the Alliance Colony; it may be that my mother's maternal relatives, the Barishes, who came to Alliance around this time, were the recipients of this act of munificence and thus the perpetuators of the story that his financial gifts benefitted them personally.

His most generous philanthropy towards settlement of agricultural/industrial communities in southern New Jersey established Woodbine, New Jersey, which I discuss in slightly greater detail later. Interestingly, although he funded schools and hospitals in the Middle-East, when Theodore Herzl asked for financial backing to found Eretz Israel, de Hirsch declined.[5] Arguments about the morality of the geo-political actions of the Israeli government's policies aside, the Baron's views about the Americas made possible the successful assimilation of most American Jews of that period and beyond. Early in the 2000s, I heard a National Public Radio program about the last of the Jewish gauchos who were dying in Buenos Aires. The cosmic connection between me and their descendants was none other than the Baron de Hirsch.

While the Am Olam movement looked to the Americas, the contemporaneous Zionist BILU movement, founded in 1882, looked to Palestine as the future home for Jews. BILU was an acronym based on a verse from Isiah (2.5), "*Beit Ya'akov Lekh Ve-nelkha*," which translates in English to "Let the House of Jacob Go!"[6] Looking west to the Americas for an assimilated and safe life for his co-religionists, the Baron viewed Zionism and the creation of a Jewish state in Palestine as fantasy. After his death, however, and into the present time, the Jewish Colonization Association, which he began, primarily funds agricultural projects in Israel.[7]

Alternate Universes Imagined

My mother's parents, children of Jewish settlers
set down near the Pine Barrens of New Jersey by the
Maurice River to farm sandy soil in cooperatives,
had escaped the pogroms' rages in the shtetls of Kiev.

New Jersey's scrub-oak land and tea-brown rivers
still ring loud and bright in my memory cells.
Years later, I learned that other lands beckoned the
settlers, strong in the imaginations of philanthropists
who made their dreams possible: Oregon, where
my ancestors thought they would have to defend the
babes from eagles after surviving the Cossacks,
and later, the Argentine, to become Jewish gauchos,
the last of them dying out in the 21st century's beginning.

The chiaroscuro of possibilities plumbs depths
of the mind's options of alternate universes,
and I reel to think of the other half-me who
might have been had my mother's grandparents
been braver or waited longer to huddle in steerage
and land, not at Castle Garden nor Ellis Island,

but at a port close to Buenos Aires, our fathers ready
to wear the dashing garb of the gauchos,
forelocks tucked into flat-brimmed hats,
britches flaring above spurred boots, or
in Oregon, tending vineyards
and shaking angry fists at hungry eagles.
From all those shades of possibility, I emerged,
shapes and lines drawn from the water and
soil of Alliance, New Jersey.

Chapter 4

Life in Alliance

1882 to 1920

My sources run the gamut: Wikipedia, Jewish and state historical on-line resources, published Ph.D. dissertations and other scholarly texts, documentary film, conversations, reprints of newspaper articles of the day, family histories and *YOVALS*—the 50th anniversary celebrations or jubilees of the founding.

Even with the generous assistance of the philanthropists and their committees, some risk taker in the old country had to have the chutzpah to scout the scene. That fell to Moses Bayuk, a member of Am Olam, a lawyer, a Talmudic scholar, and a purported intimate and card-playing crony of Leo Tolstoy in Bialystock, Russia. My mother's first cousin, I. Harry Levin, a matrilineal grandson of Bayuk, claimed that Tolstoy lost considerable amounts of money to Bayuk and another man one night. The story has been documented in different formats and passed on to my second cousin-once removed, William Levin, Bayuk's great, great grandson, but the details have blurred a little in the transmission. Allegedly, the men continued at their card game, allowing Tolstoy to recoup some of his losses, but another player, not as lucky as Tolstoy, lost big. The unfortunate, nameless loser wrote

IOUs to Bayuk and a fourth gambler, also anonymous. On their way home, probably in the wee hours of the morning, Bayuk and the other winner were picked up by the constabulary. This incident occurred shortly after the assassination of Tsar Alexander II when revolutionaries, intellectuals, and Jews were highly suspect. Somehow, Moses Bayuk's straightforward explanation of how he came to possess the IOU passed muster with the police and he was released. Not long after, he solidified plans to emigrate.[1]

Bayuk's considerable intellect, knowledge of Russian law, and success at cards did not equip him with even a modicum of agricultural knowledge. He and Eli Stavitsky, his co-scout, were fortunate that George Leach and his brother, the land-owners/lumber merchants who sold the 1,150 acres of cutover timberland in Pittsgrove Township, Salem County, New Jersey (in some accounts the original purchase was 3,000 acres), were honorable men who respected and aided the Jewish settlers. Their names, as well, were always acknowledged with respect in the family's historical references.

Bewildered and weary, yet somehow intrepid, the Jewish emigres, primarily from three Russian Jewish cities in what is now Ukraine (Kiev, Odessa and Elizabetgrad/Ylizavet-grad),[2] made their way from New York, where they had come through Castle Garden,* the main port of entry for European emigrants from 1855 to 1890. Later, Ellis Island assumed its status as the port of entry most closely associated with the Statue of Liberty and her welcome to the "tired and poor." Finally, they were delivered by rail to the wooded backwater in southern New Jersey, about five miles distant from the thriving and beautiful town of Vineland, where they would make history.

---

* "Castle Garden" is also the term used ironically by the first Alliance colonists to describe the barrack-like buildings in which they all lived until individual family dwellings were built. Throughout my life, I was confused by my family's dual use of the term until I began researching Alliance Colony history.

It is important to reiterate that although some of the early colony literature and the inscription on the gate to the cemetery state that Alliance was the first Jewish farming colony in the United States, we know that this is inaccurate. My brief summary in Chapter Two, outlining the previous attempts at establishing Jewish farming communities in the U.S., provides us with a partial list of failures. Alliance, in fact, was America's first successful Jewish farming colony. That they achieved such success is impressive; the immigrants spoke no English and possessed no knowledge of farming. Their backgrounds and experiences in the Russian Empire varied greatly. Some were highly educated intellectuals, engaged in passionate discussions about social and philosophical issues. Others were artisans who endowed the work of their hands with dignity. There were pious Jews wary of too much assimilation and those intentionally eschewing religious observance. There were those who enjoyed wealth and refinement and others who lived in downtrodden ignorance. A nearby railroad station with service to Philadelphia and New York was part of the draw for the original pioneering families who fled from the pogroms of Tsarist Russia in 1882 to settle the wild brush land at the edge of New Jersey's Pine Barrens. Even at the reasonable price of the vast acreage purchased under the auspices of the committed philanthropic agencies and the generous terms of the loans, the impoverished Jews of the Pale were only marginally better off in New Jersey. Fred Schmitt, a German farmer, was employed by the philanthropists in the first year of settlement to instruct the neophyte farmers in the basics of agriculture.

The first tasks confronting them involved cutting trees, clearing land of the tree stumps and brushy growth left after the timber was cut, and building more adequate housing than the three, long, wooden barracks, which had been constructed for the settlers by the Leach brothers, the landowners-businessmen who had sold the land to the settlers. A delegation from the

45

HEAS (Hebrew Emigrant Aid Society) established them safely, if not comfortably, in their new home, provisioned them for the foreseeable future with food and a few basics of furniture. The land was divided into fifteen acres per settler family which allowed for a house and land, every bit of which would soon be under cultivation. HEAS had purchased the land for the enterprise: one hundred fifty acres for communal usage and the remainder to be divided into sixty-six parcels of fifteen acres each. These individual parcels of land were deeded to the occupant farming families at a cost of $300 each, to be repaid in thirty-three years without interest.[3]

The first few years were difficult, and the outlook was not promising for the immigrants. After the land was cleared of tree stumps, it had to be plowed and planted. Yet within those first few years, more came. While the emigres viewed their difficult journey as their only option, it seemed to their supporters that the same fate that had befallen the other colonies would be theirs. Upon every visit by the HEAS committees, the settlers pleaded for money and assistance. The wealthy, assimilated benefactors were appalled by this importunate behavior and were not always as compassionate as they might have been.

Two newspaper articles published eight years after the original settlers arrived—"Will They Make Farmers?" from *The Sun* (August 17, 1890) and "Russian Jews as Colonists in America" from the *New York Herald* (July 26, 1891)— were printed and distributed to the attendees of the 136th celebratory picnic. Both articles exhibit a strange mix of historical reportage ranging widely in accuracy, persistent and ignorant prejudice, and compassion and admiration for the Jewish colonists. The journalists show a remarkable lack of curiosity. It's easier to imply a built-in list of tropes about The Others than to ask meaningful questions and learn the facts first-hand. On one hand, the reporters describe the clamoring pleas for money and the slavish behavior of the settlers as the typical degradation of the race. Perpetually abused people develop abject habits, such as walking in the streets rather than on the sidewalks and hanging back apologetically to be waited upon after all other customers were served in Vineland's businesses. Their protective behaviors of the old country, developed after centuries of persecution, died hard in the new country where they were still largely isolated, and these behaviors were used as justification for continuing prejudice. Once the new farmers demonstrated their hard work and agricultural successes, law-abiding and sober community behavior, their family values, humble but generous hospitality to visitors, and the high achievements of their youngsters in the area schools, the praise of their neighbors became effusive.[4]

Before farms became successful, and probably well after, other means of augmenting income were required. The first few years, children helped to earn money by walking miles to pick cranberries in the bogs surrounding the community, or they picked strawberries grown by the Gentile farmers in the area. As in Russia, some of the settlers and their children were artisans: carpenters, furniture makers, masons and blacksmiths. The farms that comprised the Alliance Community were all owned by Jewish families, and no non-Jews lived in the community until several generations later. The crews of workmen building the houses, barns and outbuildings for those of the new settlers who could afford to expand were all Jewish. No doubt the more temperate climate of southern New Jersey and the other benefits of their location soon began to yield results to those who remained healthy and motivated to work tirelessly. Although few of the Jewish immigrants had lived lives of luxury in Russia, some had been comfortable. Even those, like my great grandfather Hersh Levin the cabinet maker, who had worked with his hands, were unused to the long hours of back-breaking labor in the fields. Their surviving children became strong, healthy, competitive and successful, whether they remained on the farms or left for education and city life.

After the first few years of severe hardship, some of the farms began to succeed. The main crops in the earliest days of profitability required hours of intensive labor. The settlers planted, tended and harvested grapes, berries (strawberries, blackberries and blackcaps, also known as black raspberries which, next to blueberries, are my favorite berries of all), peaches and sweet potatoes. They also ate what they grew and relied on eggs and milk as their primary protein source. Every foot of ground not occupied by the house or outbuildings was planted, and the fields were scrupulously weeded. When crops were successful, they were shipped from the Bradway railway station (later renamed Norma) to Philadelphia and New York.

Nevertheless, the farms were not all equal, and not all the farmers enjoyed the same success. Also, despite their hard work, crops sometimes failed. It was time for a new, more inclusive and stable organization to take shape to nurture the well-being of the individual farmers and the colonies as social experiments. By March of 1883, HEAS had disbanded. As overall economic conditions in the 1890s faltered (it was, after all, the age of the Robber Barons and a period of runaway capitalism), it became uncertain what organizations were available to back the loans to the colonists. Some of the more successful farmers were paying off their farms and taking out loans to buy more land, cows and horses. With prosperity and the beginning of assimilation, some of the colonists had also begun building modest wooden houses in a modified Queen Anne style. Moses Bayuk, however, had built a more substantial two-story brick house. (In the final days of 2018, the Moses Bayuk house, in a state of severe disrepair, has been acquired by ACRe, the non-profit organization begun by Bayuk's great, great grandson, William Levin, and his wife, Malya Kurzweil Levin.)

Despite the signs of success of the farmers whose luck, skills, health, energy and persistence allowed them to prosper, some were less successful and just as persistent in asking for further assistance. Those patrons who became increasingly disillusioned by the constant entreaties for financial assistance made it very clear that there would be no handouts.

The patrons, however, could not in good conscience sentence the immigrants, who had already been through so much hardship, to life in the over-crowded, insalubrious tenements of the lower-east side of New York. To accommodate those of the settlers who were not well suited to farming, the wealthy backers endeavored to establish light industry, specifically a cigar factory in the earliest days. Through such assistance, these philanthropists who were assimilated, westernized, financially successful, and business oriented, had sway. They made enormous investments in aiding their Eastern European co-religionists, and while they could not draw up and establish by-laws, their influence—in the form of promising further investment in the community or threatening to switch their support to other enterprises—was no doubt felt.

Expulsion of "undesirables" or those who complained about conditions was a method of control exercised by HEAS. Since the beginning of the colony, AIU and HEAS—the capitalist leaders—promoted industry as much as agriculture and worked to discourage the agrarian, utopian and socialist leanings of the colonists. In February 1883, seventeen families were accused of laziness and banished to a short-lived colony named Estellville, which was established by unscrupulous land developers in an infertile area near Mays Landing, New Jersey, about twenty miles distant. Whether the original accusations were first voiced by other colonists or the backers is unclear. It seems that these charges were lodged by the New York industrialists who had little first-hand knowledge of life in the colony. At any rate, fifteen of the seventeen returned to Alliance in dire straits and occupied the old

"Castle Garden" barracks. When the HEAS secretary came to settle the dispute, he was taken captive and forced to listen to the pitiful cries of the famished children begging for bread. The ring leaders were arrested but found innocent of assault and released. Their champion was Augustus Seeman, the New Jersey commissioner of immigration.[5]

It occurs to me that rather than lazy, some of them may have possessed other skills than the ability to work tirelessly at physical labor. Perhaps among them were those who listened deeply to and empathized with their family or community members who were suffering; perhaps their skills were in nurturing and they may have been successful educators or social workers, given half a chance. Maybe some of them saw no point in working their fingers to the bone with no energy left to observe beauty around them or to find inspiration for making art, creating music and telling stories. In cultures or situations that demand ceaseless work, there is little tolerance for those destined to be caretakers, musicians, artists and storytellers.

In 1900 in New York, the Jewish Agricultural and Industrial Aid Society was organized as a subsidiary of the Baron de Hirsch Fund to provide several benefits to the agrarian immigrants from Eastern Europe. The organization extended loans on generous terms for cooperatives as well as individual farmers, with an emphasis on self-supporting agricultural activities. They established small rural industries to provide supplemental incomes to the colonists, and their extension agents fostered innovative, hands-on agricultural training opportunities. *The Jewish Farmer*, a Yiddish-English monthly publication offered advice, and the Bureau of Educational Activities met some of the cultural needs of the colonists. Unlike some of the prior organizations that discouraged the inexperienced from farming, the mission of their Industrial Removal Office, which became autonomous after 1907, supported relocation of urban immigrants from cities. After

World War II, their diversified programs extended to thousands of displaced persons; this work kept the community of Alliance and others viable with the establishment of kosher poultry farms.[6]

ᴏᴠ

While the Alliance Community never, to my knowledge, drew up a set of by-laws as did other colonies (New Odessa, for example), neither did they ever elect an official leader. In 1891 there were 612 persons who comprised the colony, and yet there was no one acknowledged as a leader and no system that could be called a government. The unnamed author of the article "Will They Make Farmers?" originally printed August 17, 1890, in *The Sun* newspaper (and reprinted by the Jewish magazine *The Jewish Farmer*), considered that the settlers lived in "the purest and most ideal state of anarchy"; however, as he expressed this sentiment, he also doubted that few, if any, were familiar with the political system of anarchy.[7] Apparently, he was totally unaware of the extent to which some of the colonists had benefitted from the secular education they were able to access in Russia. He could not have known of the Am Olam movement nor the idealistic commitments of some of the colonists who believed in the communalistic and democratic, socialistic vision evinced in prior attempts at colonization elsewhere in the country. Had he possessed the slightest inkling of the depth of their intellectual abilities and propensities towards ardent, philosophical beliefs and heated discussion, he never would have written such a statement.

Although no laws nor officials governed the Colony, there was a Justice of the Peace in Centerton and another in Elmer, villages near the Jewish community. The *Sun* article states that in the eight years of settlement, there had never been an occasion when it became necessary to call in the authorities.

There was no drunkenness that erupted in a disturbance of the peace, no serious brawling in which injuries were sustained.[8] Apparently, the reporter was unaware that Jews mostly used alcohol in the form of wine for religious ceremonies rather than for socializing and dissolving the inhibitions that help to enforce social norms. It was customary for the adult men to meet every two weeks in the meeting rooms of one of the synagogues and to discuss the affairs of the colony. Each had his say, and all listened. The fraternal spirit that existed naturally developed so that each proposal considered in these meetings was voted upon, with the majority prevailing. Conscience must have tempered any heady over-reaction to their newfound freedom because this system worked. If a man behaved dishonorably, he was shunned. Just as the reporter was ignorant of the educational level of some of the settlers, he must also have been ignorant of the expulsion of the seventeen families in February of 1883 to the ill-fated Estellville Colony. Certainly, even seven and eight years after its occurrence, this episode could not have been widely broadcast by the colonists who must have been eager to forget the painful event and the negative press it received.

In contrast to the seeming lack of governmental structure in the Alliance Colony is the example of Woodbine. As a great grandchild of Alliance Colony founders, I absorbed much of the history in which my mother's family had been steeped. Therefore, I always knew something about or someone whose roots in the new country went back to the colony of Woodbine, which was about thirty miles distant from Alliance. Established in 1891 with the intention of making it an agricultural/industrial colony from its inception, Woodbine was the largest of New Jersey's intentional Jewish communities. Whereas Alliance and the other earlier colonies were mostly initiated by the colonists, the plans for Woodbine were deliberately conceived and carried out by the sponsors. As the Baron de Hirsch's negotiations with the Russian government to establish safe communities for the Jews in the Russian Empire failed and the pogroms increased in vehemence and frequency, de Hirsch saw the need to create a carefully planned community in New Jersey. Woodbine became the child of the Baron de Hirsch Fund, a direct application of the principles of the foundation.[9]

In 1903, it was incorporated as a borough in which all official posts were held by Jews, making it not just the first town in America to hold this distinction, but the first self-governing Jewish community since before the Diaspora of Jews in the biblical era.[10] Although it closed in 1917, the reputation of the Baron de Hirsch Agricultural School's training programs in affecting positive change in New Jersey's farming practices was well-known. Even in my childhood I possessed a sense of Woodbine's historic significance. Woodbine, despite the considerable planning and resources that went into its establishment, existed overall for a much shorter period than did the Alliance communities. Academic sources on the subject document and evaluate the successes and failures of Woodbine as well as the other colonies and provide more information for anyone seeking a more thorough understanding.

The original Alliance Colony of forty-three families, founded in May of 1882, was augmented with successive groups well into the 1920s, as the pogroms continued in the Russian Pale. My mother's mother came as a teenager with her parents and siblings in the second or third group to be brought to the colony. In these next waves of immigrants were some who had studied in fields associated with agriculture. They and the various funding organizations introduced new crops, such as asparagus and white potatoes. Light industry and focused agricultural education furthered the development of the expanding communities.

ҩ

Industry Comes to the Alliance Colony

Very few of the farmers were self-sufficient agriculturalists and resorted, almost immediately, to light manufacturing to augment their incomes, especially during the winter when nothing grew. In the first winter of the Alliance settlement, two of the three large barracks initially used to house the settlers were allocated for light industry: cigar making in one and sewing in the other. The cigar factory burned down in less than a year, possibly through an act of arson perpetrated by a laborer who was disgruntled over a salary cut. It was rebuilt but closed shortly thereafter.[11]

Eternal class struggles that can still divide citizens of a country or region were present in the social fabric then: farmer against factory worker, worker against capitalist, foreign laborer against Americanized businessman. With the socialistic/idealistic background of Am Olam pitted against the necessity of making a living, strife followed the establishment of these factories. The Jewish newspapers of Philadelphia and New York exposed the exploitive wages and working conditions of some of the short-lived factories.[12] In Alliance where the farms and houses were on the same piece of land and the factories a short walk from the home, the symbiotic balance between summer farming and winter manufacture was logistically easier than in the communities of Carmel and Rosenhayn where the farms and homes were at a greater distance from the factories. Many more residents of Carmel and Rosenhayn were dependent upon their sewing machines, and employment as machine operators, cloak makers and tailors; consequently, they farmed less. The sewing machines became both bane and blessing.[13]

Despite criticism, by some, of the encroaching but necessary industrial expansion, it took place, stimulated by private and organizational efforts. Cigar manufactory was short lived in the colony. The industrial endeavor to have the most significant impact upon the colony was that of Brotman and Company. Abraham Brotman was a recent immigrant from Galicia who moved his successful clothing factory from New York to a small piece of land at the northern end of Alliance where he provided houses and employment, funded in large part by another Baron de Hirsch ad hoc foundation. The factory became the center of the primarily industrial community. In 1908, the village—really a company town, albeit a benevolent one—numbered two hundred people who lived in homes owned by the Jewish Agricultural and Industrial Aid Society. The Brotman family lived locally but ceded their commercial property rights to JAIAS.[14]

Eponymous Brotmanville, which in its earliest days was known as Brotmansville and still, in 1982, referred to as such by some of the last first-generation children of the settlement, exists today as an African-American community; the old synagogue is now Mount Moriah Baptist Church. Rich Brotman, great grandson of Abraham Brotman and documentary film maker, interviewed his father and others for his 1982 film *First Chapter in A New Book, The Story of Brotmanville and the Alliance Colonies*. During the interview, Rich's father, Stanley, a New Jersey judge, turned the tables on his son, the filmmaker, and asked what he thought of Brotmanville. Rich explains that when he first learned about the factory established in Brotmanville, he was concerned that his great grandfather may have been an exploitative capitalist running a sweat shop.[15] For anyone familiar with the history of Jewish immigration to the Lower East Side of New York in that era and the struggle between unionized laborers and capitalist factory owners, the tragedy of the Triangle Shirt Factory is the first event that comes to mind. That Abraham Brotman supplied housing for his factory workers and did not abandon them when contracts were completed shows his merit, and

Rich Brotman's relief that his great grandfather cared about connection and community touches me.

In my research on the history of the Alliance Colony, I have learned something about my grandfather that fills me with the same sense of rightness and pride that Rich realized about his great grandfather. It is something about which I had no familial clues, something totally lacking in family lore—or at least in my memory. Whether John Levin's participation in a 1913 labor strike was a skeleton in the closet that no one talked about or something my young consciousness did not encompass, I now have no way of knowing. In October of 1913, my mother was almost three years old; I don't know whether my Poppy and Nanny were farming or whether he was working in the light industries. This bit of family history is gone to me forever, but apparently John Levin and my grandmother's brothers, Nathan and Samuel Barish, who at least a generation later had become entrepreneurs in the manufacture of women's clothing, were among eleven long-time colonists who supported workers striking at a Brotmanville factory, by then owned by Philadelphia's Kramer and Sons. In Rosenhayn, only five or so miles from Alliance, a 1900 strike led to the formation of a local chapter of the United Garment Workers' Union. Local newspapers claimed that outside agitators instigated the 1913 strikes and riots, but the names of the young Alliance men in the Salem County court case against the "rioters" indicate a social consciousness and local support for the labor rights of their neighbors. I feel pride in this young man who stood up for workers' rights and am satisfied with their acquittal by a jury of their peers in the court cases brought against them by the factory owners.[16]

In addition to the brief incarnations of the cigar factory and the various garment factories which provided home-based piece work for the winter employment of the farmers, in 1901 Maurice Fels, of the Philadelphia Fels Naptha Soap Company, established a canning company in Norma. Fels, with the cooperation of the JAIAS, built the Allivine Cannery (a combination of the Alliance Colony and the town of Vineland) to afford a ready market for various produce and to augment the quantity and quality of the farm products they raised. Sidney Bailey, an ardent Am Olamer, who arrived in Alliance the year following the original forty-three, explains how asparagus, white potatoes, cherries and apples soon expanded the standard crops of grapes, berries, sweet potatoes and peaches, all of which found their way to the cannery as well as fresh markets.[17] Fels was a generous supporter of a variety of rural educational initiatives and subsidized many individual teachers. The cannery paid fair prices for the berries and vegetables the farmers raised, sold fertilizer to the growers at cost, and provided them with assistance in improving crop yields. The canning factory had a reputation for cleanliness, efficiency and good working conditions. It was sold between 1916 and 1917 to the Torsch Packing Company.[18]

Until the 1890s, when plant diseases and other problems caused the Welch Grape Juice Company to move to New York State, that company had also provided a market for local grapes. Dr. T. B. Welch, a Vineland dentist and teetotaling Methodist, was strictly opposed to alcoholic beverages, even in the form of communal wine for religious ritual. His business was succeeded by the Vineland Grape Juice Company and eventually became an experimental station for viticulture run by the U.S. Department of Agriculture.[19]

∾

The Cartography of The Alliance Colony

"Have you thought about moving to town or to another area of the Ozarks where you would have more friends?" This question, so often posed to us by other people, is in the back of our minds as my husband and I age on our forty acres, in good health but somewhat isolated from our community of friends. We have a younger neighbor/friend who is caring and reliable, and we know that we can count on her and her husband in semi-emergencies that would not necessitate an ambulance call. Across the river, about five miles away, is a more concentrated community of back to the landers, longtime friends, and they urge us to buy a small piece of land near them and augment the community with our energy. That option has its appeal, but it is unlikely that we can afford what we would want in a piece of land and house. While we still have our health and strength, we seem unable to cut ourselves off from this small plot of land we have cultivated over the years. If life brings changes that make us unable to garden, we will both feel that a part of us has been amputated and will, the rest of our years, feel the phantom pain of the missing gardens that we have tended for nearly forty-five years.

The other option in the pondered question is that of moving to town. Some of our long time back-to-the-land friends have moved "back from the land"; most of these people lived closer to Fayetteville and have found in it an attractive college town/small city atmosphere. Fayetteville real estate is prohibitively expensive for us and out of the question. The town of Marshall and our mailing address would never be under consideration while Leslie might be but for our absolute opposition to living in a small community where everyone thinks they know everything about you. The lack of cultural, ethnic, racial and religious diversity is one of the drawbacks of Ozark towns and the region in general; it is one of the aspects of life here that has never satisfied despite the many positive qualities.

And yet throughout this memoir, I have romanticized the experiences of my childhood and teen summers in a string of small hamlets connected by a narrow set of common attributes. While I reject a small Ozark town because of confining social norms, in my mind, I still go back to the rural, Jewish, South Jersey community of my childhood and youth and view it through rose-colored glasses. I am not immune to the lure of a mythological Eden—of an American, agrarian childhood at the grandparents' farm which, when painted in popular culture with broadly nostalgic brush strokes, I almost always deem too sentimental and saccharine.

So, here is a graphic representation of the streets and buildings of the tiny communities of Alliance, Norma and Brotmanville as they looked in 1901, well before my time but similar enough and characterized by familiar names. Although the tiny hexagonal community of Six Points—just due west of Brotmanville, where Willow Grove Road/(Parvin) State Park Road, Garden Road and Rainbow Lake Road/Six Point Road/Alvine Road formed an interstice—was considered a part of the Alliance Colony, it is not on this representation of a hand-drawn map. Upon first seeing this map, so many remnant memories flooded back into my consciousness that I felt a palpable sense of "being there." I could have driven the rental car to all the places I knew from childhood without the benefit of GPS. It is a map depicting the Alliance Colony at the turn of the 19th to 20th century, and it positions much history of the founders of this small community.

Jay Greenblatt, the president of the Alliance Colony Foundation, told me via email in the autumn of 2018 that the map is a condensed, printed version of the hand-drawn map done in the summer of 1982 by two surviving matriarchs of the colony, "the result of driving those two old ladies (Mollie Greenblatt Kravitz and Elizabeth Colen) around so they could argue with each other until they agreed on each item." Jay thought that his Aunt Mollie was the one who drew the map. The map was distributed with the packet of materials to the attendees at the 2018 Alliance picnic.

It depicts an area of land approximately two miles from north to south and one mile east to west. The main road

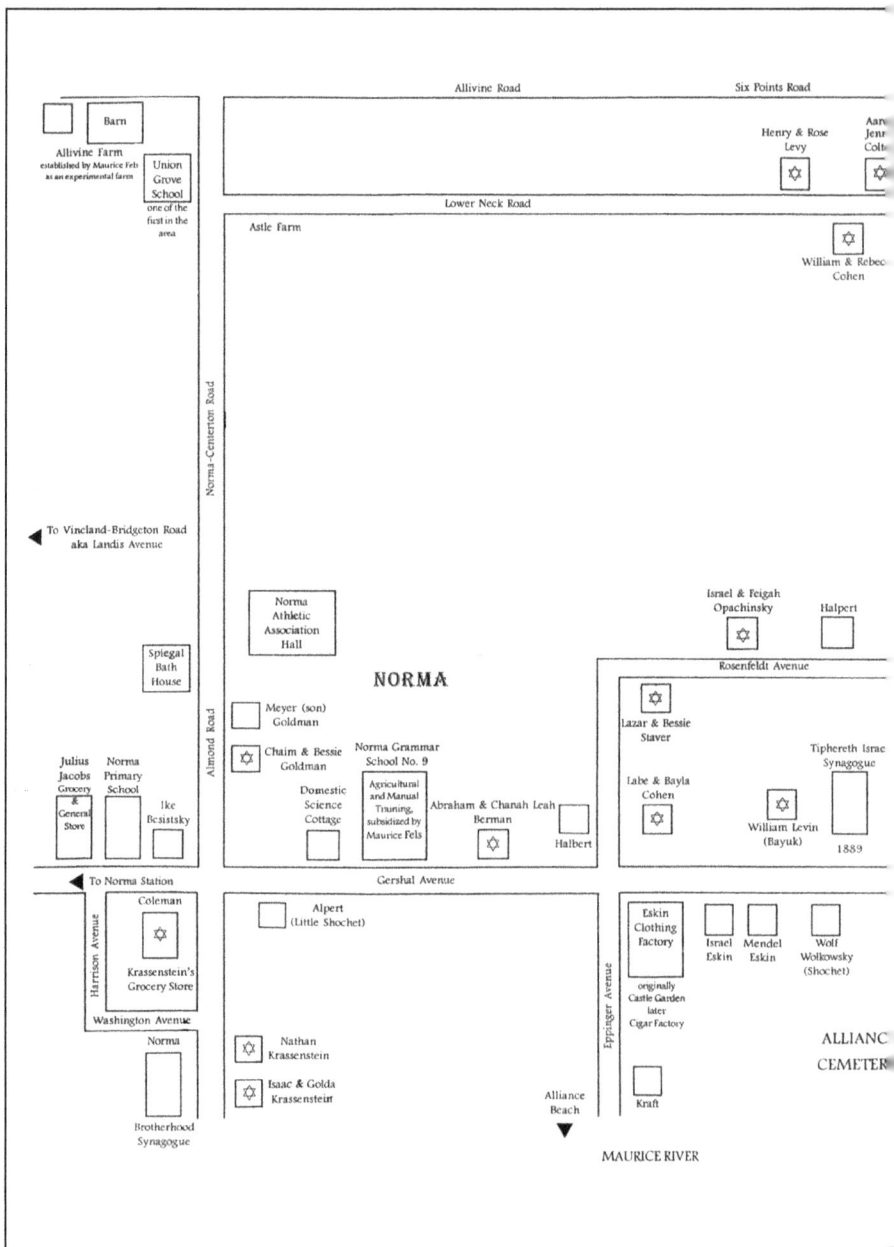

Allivine Road

Six Points Road

Barn

Henry & Rose
Levy

Aar[…]
Jenn[…]
Colt[…]

Allivine Farm
established by Maurice Fels
as an experimental farm

Union
Grove
School
one of the
first in the
area

Lower Neck Road

Astle Farm

William & Rebec[…]
Cohen

Norma-Centerton Road

To Vineland-Bridgeton Road
aka Landis Avenue

Norma
Athletic
Association
Hall

Israel & Feigah
Opachinsky

Halpert

Spiegal
Bath
House

Rosenfeldt Avenue

NORMA

Almond Road

Meyer (son)
Goldman

Lazar & Bessie
Staver

Tiphereth Israe[…]
Synagogue

Julius
Jacobs
Grocery
&
General
Store

Norma
Primary
School

Chaim & Bessie
Goldman

Norma Grammar
School No. 9

Labe & Bayla
Cohen

Ike
Besistky

Domestic
Science
Cottage

Agricultural
and Manual
Training,
subsidized by
Maurice Fels

Abraham & Chanah Leah
Berman

William Levin
(Bayuk)

1889

Halbert

To Norma Station

Gershal Avenue

Coleman

Harrison Avenue

Alpert
(Little Shochet)

Eskin
Clothing
Factory

Israel
Eskin

Mendel
Eskin

Wolf
Wolkowsky
(Shochet)

originally
Castle Garden
later
Cigar Factory

ALLIANC[…]
CEMETER[…]

Krassenstein's
Grocery Store

Washington Avenue

Norma

Nathan
Krassenstein

Isaac & Golda
Krassenstein

Alliance
Beach

Kraft

Brotherhood
Synagogue

MAURICE RIVER

Eppinger Avenue

60

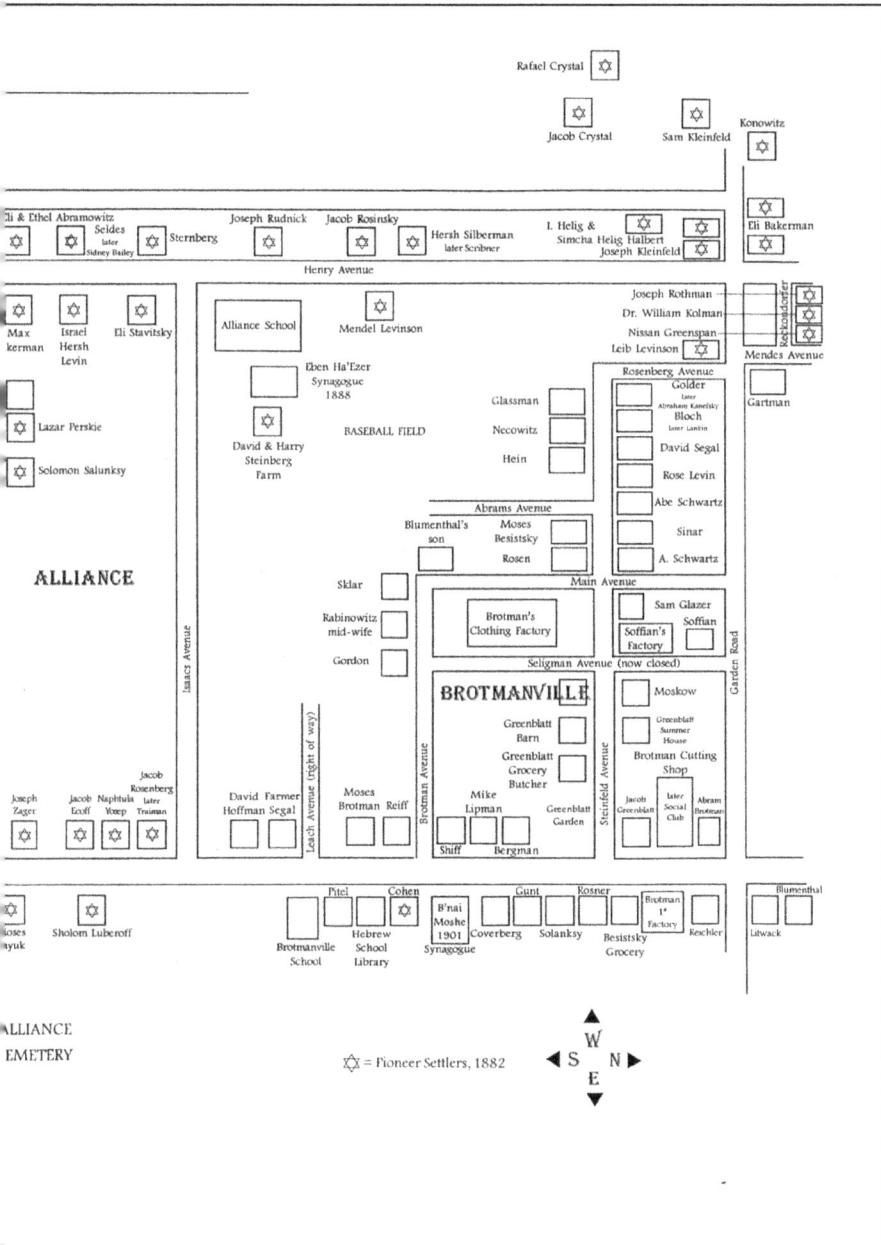

Rafael Crystal ☆

Jacob Crystal ☆   Sam Kleinfeld ☆   Konowitz ☆

Eli & Ethel Abramowitz ☆ Seldes ☆ later Sidney Bailey Sternberg ☆   Joseph Rudnick ☆   Jacob Rosinsky ☆ ☆   Hersh Silberman later Scribner   I. Helig & Simcha Helig Halbert ☆   ☆   ☆ Joseph Kleinfeld   Eli Bakerman ☆

Henry Avenue

Max ☆ kerman   Israel ☆ Hersh Levin   Eli Stavitsky   Alliance School   ☆ Mendel Levinson

Joseph Rothman ☆
Dr. William Kolman
Nissan Greenspan ☆
Leib Levinson ☆   Mendes Avenue

Eben Ha'Ezer Synagogue 1888   Rosenberg Avenue
Golder later Abraham Kanefsky Bloch later Lankin   Gartman
Glassman

☆ Lazar Perskie   ☆ David & Harry Steinberg Farm   BASEBALL FIELD   Necowitz
Hein   David Segal

☆ Solomon Salunksy   Rose Levin

Abrams Avenue
Blumenthal's son   Moses Besistsky   Abe Schwartz

ALLIANCE   Rosen   Sinar
A. Schwartz

Sklar   Main Avenue
Sam Glazer
Rabinowitz mid-wife   Brotman's Clothing Factory   Soffian's Factory   Soffian

Gordon   Seligman Avenue (now closed)

BROTMANVILLE   Moskow

Greenblatt Barn   Greenblatt Summer House
Greenblatt Grocery Butcher   Brotman Cutting Shop

Joseph ☆ Zager   Jacob Naphtula ☆ Ecoff   Jacob Rosenberg later Vorep Trusman ☆ ☆   David Farmer Hoffman Segal   Moses Brotman Reiff   Mike Lipman   Greenblatt Garden   Jacob Greenblatt   later Social Club   Abram Brotman

Shiff   Bergman

☆ oses ayuk   ☆ Sholom Luberoff   Pitel   Cohen ☆   B'nai Moshe 1901 Synagogue   Gunt   Rosner   Brotman 1ª Factory   Blumenthal
Brotmanville School   Hebrew School Library   Coverberg   Solanksy   Besistsky Grocery   Reichler   Litwack

ALLIANCE
EMETERY

☆ = Pioneer Settlers, 1882

Isaacs Avenue
Leach Avenue (right of way)
Brotman Avenue
Steinfeld Avenue
Garden Road

61

north to south, Gershal Avenue, is where John H. Levin, my grandfather, lived during the first two decades of my life, but this map does not extend the whole length to include that part. His house, far south on Gershal, seemed to be at least a couple miles from the beach on the Maurice River. His stretch of the road was not part of the historic Alliance settlement but consisted of—for about another mile—houses, woods and farmland. Many of the houses immediately south of Landis Avenue, which runs east-west through the entirety of the city of Vineland, were bought as early as the mid-1920s through the mid- to late-1940s. Some were purchased in the 1930s by families of German Jews who wisely interpreted political trends in Germany, as Hitler started gaining power, and got out in time. Post-war Holocaust refugees from farther east in Europe came to the Jewish farming community in the next wave of immigration and had to live at the far reaches of the settlement where land was still available. This area was also inhabited by the next generation of the settlers' families.

At either end of its length, Gershal Avenue is named Jesse's Bridge Road, after the bridge that spans the Muddy Run about a mile south of what was my grandfather's property. Jay Greenblatt tells me that at some unspecified time in recent history, Pittsgrove Township officials wanted to change the Gershal Avenue section to Jesse's Bridge Road for consistency. Fortunately, Jay prevailed, and the history of the Alliance Colony was powerful enough to preserve the Gershal name throughout the area that was founded as the Alliance Colony. My sense of the distances between places I once knew so well is measured in childhood miles. I did not get a driver's license until I was in my late twenties, so I never drove when I visited there as a teenager, and anticipation always colored my perception of distances and the time it took to arrive at the desired destination. Even now, at seventy-eight, given good walking shoes, appropriate clothing, and water, I could comfortably walk the whole length of Gershal Avenue.

Of the eight roads intersecting Gershal Avenue in an east-west direction only two completely exit the community. One is Almond Road at the southern end of the map and the other is Garden Road at the northern end. An arrow near the left edge of the page points to a main east-west road, Landis Avenue, the road that connects the towns of Vineland and Bridgeton.

The smaller roads chopping up the farmlands and woodlots were named after the New York luminaries and philanthropists who bankrolled the various funds that established the original colony. Gershal, itself, is a changed spelling of the name of one of New York's prominent Jewish leaders from the era, Leopold Gershel. Eppinger Avenue, only a couple blocks long, dead ended at "The Beach," on the Maurice River. Known as the Alliance or Norma Beach, the land was privately owned and functioned first as the community swimming hole from the time of the first settlers, and then as a country resort for almost a century before time wrought its inevitable changes. Eppinger, Shiff (Schiff), Isaacs, Henry, Rosenfeldt, Abrams, Brotman, Steinfeld, Seligman and Rosenberg were through streets or dead ends into farmers' fields and were all named after the New York benefactors. As the community developed, shops and places of worship were established to meet the basic needs of the Jewish farmers. Two shochets, or kosher butchers, and several grocery stores were opened along the eastern length of the colony from Brotmanville to Norma. Norma boasted the Spiegal Bath House, which must have served the entire community, on Almond Road west of Gershal Avenue.

In Alliance proper in 1901, two synagogues existed. Eben Ha'Ezer Synagogue (1888), the more reform in observance, was adjacent to the Alliance School. This shul (Yiddish for synagogue) had a social hall which hosted cultural events as well as religious services. Commanding a bit of a view, Tiphereth Israel, the other synagogue, was built in 1889 on the slightest rise of land that one could scarcely call a hill. On the eastside of Gershal Avenue the B'nai Moshe Synagogue was built in 1901 at the southern edge of Brotmanville. Norma, at the other end, had the Brotherhood Synagogue which was incorporated in 1912. The Alliance Cemetery was established east of Gershal Avenue and north of Eppinger Avenue. In 1901, according to the map, the Eskin clothing factory, Eskin family houses, and several others, including the Moses Bayuk house, lined the road and the cemetery was set well back. These days the crumbling Bayuk house stands, but the expanding cemetery lawn comes right up to the road. The *Chevra Kadisha* or burial/cemetery society was established in 1882 and is still a strong organizing entity for the Alliance Community.

In the vague area that is neither clearly delineated as Alliance nor Brotmanville, though "Alliance" is the umbrella term that includes Brotmanville and Norma, and is marked by houses of people who came after the original settlers, one house is identified as the dwelling of "Rabinowitz mid-wife." Although her first name had been forgotten by the mapmakers in 1982—two women who may even have been delivered by her hands—Mrs. Rabinowitz occupied a position of great importance in the whole Alliance Colony. Mrs. Rabinowitz quite likely delivered my mother in Norma on December 26, 1908, in that small post office. It is amazes me that both Norma and Alliance—what I think of as those two tiny "sub-divisions" of the Alliance Colony proper—possessed post offices. The post office locations are not marked on this map. Various qualified members of the community served as postmaster or postmistress for many years. According to additional family history written by Bea Coltun Harrison, my mother's cousin, her father Aaron Coltun was appointed the postmaster of the Norma post office in 1906. The post office and general store were in the front of the building with their living quarters in the rear. They possessed the first telephone in Norma, but whether Mrs. Rabinowitz had a phone as well is the algebraic X in the equation. At any rate, someone summoned her, and perhaps this uncle through marriage was on duty when the local midwife presided at my mother's birth in his bailiwick. Nothing remains except my memory of mother's statement that she was born in the post office and her pleasure at my childish joke that she was a special delivery.

The BASEBALL FIELD was a stone's throw from where my Pop grew up. It seems significant to me that the mapmaker, Mollie Greenblatt Kravitz, capitalized this place on the map since the village names are the only other geographic locations that appear in all caps on the map. The importance of baseball in Mollie's young life was mirrored by the residents, as well as visitors to the community. In his unpublished, handwritten memoir (July 29, 1933) "Colonial Characters," Louis Mounier, a visiting educator, praises the game and the Norma Athletic Association squad, which included Alliance and Brotmanville. The baseball team was the major activity of the association. Herman Eisenberg, brought up on a farm in Norma, describes the significance ascribed to the team:

> That, more than anything else, gripped our youthful imagination, which showed more than anything else that we were really Americans. The original team was uniformed in knickers and golf hose, the official bicycle uniform of the day. Every member of the team was literally worshipped by the youngsters. Connie Mack, Eddie Collins and Ty Cobb may have had their own admirers, but first in our hearts ranked "Jake" Dittis, "Jake" Spiegel, George Beebe and Bill Beebe and the other members of that gallant team: Ben Dorshow, John Levin, Moe Spiegel and others. No varsity letter was more eagerly coveted than the "N.A.Λ." and no varsity athlete was more envied and admired than a member of

the team. Victory for the team on Saturday afternoon was a veritable holiday. Defeat left us all in gloom.[20]

The next team (circa 1910) was lucky enough to get real baseball uniforms. Their ardent fans cheered their plays and professional look from a newly built grandstand. The baseball field eventually made way for an asparagus patch and cornfield, but the glory of those early teams lived on in the hearts of the players and their fans for years. In 2018 while living in British Columbia, Rita Shreiber, daughter of Rosie, a lifelong friend of my mother, proudly posted on Facebook a photo of the 1910 team, on which her grandfather, Raymond Shreiber, played. Other descendants quickly identified their grandfathers, as well, while I vainly sought for mine, who had only been on the first team.

The other couple houses of original settlers and twelve houses of later settlers were strung out along the roads. The rest of this odd-shaped northern section of Alliance was farmland. The long narrow strip of central Alliance had three dwellings on Henry Avenue in the west. My great grand-parents' home, the place where my grandfather was born, was between two other families of the forty-three, two along Shiff Avenue and four along Gershal. The southernmost section of Alliance, cut from the northeast corner of Norma, housed three original settler families and the synagogue Tiphereth (also T'ferith) Israel.

Tiphereth Israel, the more orthodox of the synagogues, was constructed in 1889 and still stands, one of the few remaining vernacular synagogues of that era in the country. Howard Jaffe, about sixteen years younger than I, but like me, a summer-kid—from Newark, New Jersey—is a descendant of an early family and is married to Deborah Jaffe, whose family came as Holocaust refugees. Howard has long resided and farmed in the area. His enormous dedication and effort have kept the building in good repair. He took me into the building on August 11, 2018, where Susan Kehnemui Donnelly, also a descendant of the original settlers, was interviewing others for her documentary film, in progress, *Alliance*. I am fortunate that she asked to interview me. Although I am not an observant Jew, sitting on the bimah, the podium in front of the Ark where the reading of the Torah takes place at religious services—the heart of synagogue in a sense—and talking about my memories of Alliance and Norma was thrilling. These days, Jewish holidays and cultural activities are observed and celebrated in Tiphereth Israel, due in part to the work of my cousin William Levin, his wife and co-director of Alliance Colony Reboot or ACRe, and the community that they are building. Next to the synagogue still stands the house that was once the home of his great grandparents, William and Lena (Bayuk) Levin, my great Uncle Bill and Aunt Lena. As a child, I loved to visit them. When my husband and I upgraded our very rustic kitchen of rough-cut oak shelves and counters to simple cabinets built by a friend, I tried to capture the spirit, if not the exact look, of my Aunt Lena's white-cupboarded, country kitchen. Two other houses of the first settlers marked the corners of that section of the map, the rest constituting cropland.

The Norma section of the Alliance Colony is still larger than Brotmanville and on the map was mostly indicated as farmland. All but one of the houses occupied by the colonists of 1882 were clustered along the roads far to the east with only one tucked into the far northwestern corner. Even so the home of William and Rebecca Cohen was not isolated. Across Lower Neck Road were two neighbors, and another cluster of original settlers lived across Shiff Avenue. Also in Norma was the Norma Athletic Association and the Norma Grammar School No. 8—an agricultural and manual training institution/workshop for boys and a domestic science cottage for girls—subsidized by Maurice Fels, the Philadelphia industrialist and philanthropist of the Fels Naptha Soap Company. Fels, a bachelor with a sizable fortune, was supportive of education, especially practical, trade-oriented education to improve under-educated communities.[21]

Early in the Colony's history, the natives of Vineland had considered Norma the "dirtiest of 'Jew towns'."[22] This term of derision was also used for the Alliance and Brotmanville sections of the Alliance Colony. In Brotman's documentary, Jake Helig, the last of the farmers of my mother's generation and a wonderfully colorful character, chews on his cigar and explains the frequency with which this antisemitic slur was used in the early days until some of the strong, young Jewish men defended themselves and their communities against the nativist prejudices with their fists and put an end to the insulting term.[23] The investments in the communities by Maurice Fels and others did much to improve the hamlets and dissolve some of the prejudices.

Fourteen buildings constituted the eastern most portion of Brotmanville, nine of which were houses, only one belonging to an original settler. The public school, B'nai Moshe Synagogue of 1901, and the Hebrew School Library stood close together at the south end. A grocery store and the first Brotman factory completed this part of the settlement on the east side of Gershal Avenue. The buildings snuggled together with farmland behind them but not between them. The first block of land settled by the Brotmans was bound by Brotman, Seligman, Steinfeld and Gershal Avenues. Three family dwellings, none belonging to original settlers, were included. The Greenblatt family butcher shop, barn, garden and some farmland were encompassed in this block with the main Brotman Clothing Factory, post-dating the smaller one across Gershal, in the lot due west. To the north a lot with two dwellings and the Soffian factory abutted Garden Road. Across the short stretch of Seligman Avenue, which is now closed, three more dwellings and the Greenblatt "Summer House" completed the more settled area of Brotmanville.

As a child I thought there was a mysterious, but unidentifiable geographical mark delineating Norma, Alliance and Brotmanville, but they were all part of the Alliance Colony. Nevertheless, to this day my second cousins who grew up in the country distinguish between Norma and Alliance as the places where they were raised, though only half a mile or less separates them. The Bradway Station changed its name to the Norma Station because somewhere else in New Jersey there existed another Bradway Post Office. The story goes that the postmaster's daughter was named Norma, and from that time onwards that area of the Alliance Colony was known as Norma. On the map, the Alliance Cemetery and the Maurice River were designated simply with capital letters but no symbols. No semiotics such as parallel wavy lines signified the river in its directional flow. No clusters of gravestones with *Mogen David*, or Jewish stars, depicted the cemetery; these were reserved for the houses of the original forty-three families. West and south of the Brotmanville settlement, the map shows a greater concentration of homes belonging to the original settlers. The lines between the sections of north Alliance and Brotmanville blur; farmhouses had more open land surrounding them and, one hopes, wooded areas, too.

∾

Early Memories

The 1932 *YOVAL* features seven short remembrances by the more educated children of the settlers. They were my grandfather's peers. While acknowledging that their lives were lacking in luxury, the writers of these commemorative accounts were conscious of the progress made and the comforts secured by those who remained in the community. One and all declared the same belief, despite opposing political differences, levels of education, and degrees of religious observance—that they had lived a simple, natural life.[24]

The four languages in use in the Alliance Colony at the beginning were Russian, German, Hebrew for religious worship and observance, and Yiddish—the low German dialect written in Hebrew letters and spoken as a lingua franca by Eastern-European Jews. There were no night classes in English in rural southern New Jersey in the late 1880s, so adults learned English from their children who went to the public schools and the smatterings they picked up through their interactions with the locals in the outside community. The first-generation American children grew up speaking English without accents while retaining their ability to communicate in Yiddish.[25] My parents' facility with Yiddish was somewhat diminished from that of their parents. While they probably would not have understood the entire dialogue and the subtleties thereof at a Yiddish theater production, they were fluent enough to employ the language to discuss issues in front of my brother and me while concealing the subject under discussion from our ever-alert attention. My brother and I, of course, knew the words and expressions in common use, many of which have since become a part of contemporary English.

The *YOVAL* contributors recalled theatrics and lectures by visiting dramatic companies and professors, organizations that they formed, romances that blossomed. They called it their "Golden Age," a term I sometimes apply to my childhood and teen years in the penumbral time and space I spent in the colony. I use the term "penumbral" because my time there was not the source, not the original substance, but a removal several times from that solid history and already a fading shadow.

My grandfather was one of the first American-born generation in this settlement of *shtetl* Jews transformed into cooperative farmers in the new world, certainly the first new-world child for his family. Before I was born in May of 1941, my grandfather had moved back to Norma to "five acres and independence" after a span of intervening years as a small-scale merchant in Atlantic City. Fifty-nine years after the settlers came to the area of Pittsgrove Township, which I knew as Alliance and Norma, New Jersey, I was the second child of my parents, Sylvia and Meyer Weinstein, born in Philadelphia, some forty miles distant from the Alliance Colony.

I was born into a family in the extended ripples of this small but nurturing pond of ethnic and religious homogeneity. It was not a re-creation of the ghettoized communities from which they had come, but rather a grand cultural experiment to make the despised and beleaguered Jews of Eastern Europe something they had not been in centuries—agriculturalists. At some point after my mother, and possibly her middle sister, were born, John and Bessie Levin moved to Atlantic City, New Jersey. In the middle to late 1930s, they left their life as small-scale merchants in Atlantic City and returned to the country. By the time I was old enough to have a consciousness of the place, my grandmother had had a stroke and was in a convalescent or nursing home in Vineland. I only knew my grandparents' lives in Atlantic City from photographs and stories, mostly of my mother's life as a teenager and young adult. My parents' wedding photos that I had so treasured burned in our house fire in 1976, and I was never able to find duplicates in the jumbled boxes of photos belonging to my brother and sister-in-law. Of my grandfather, there are many stories, but even more memories of activities, events, things we did. Just as vivid are the spatial sense memories of his house, the furnishings, the appliances and utensils in the kitchen, the close to one hundred trees he planted in the years that he lived on his five acres on Gershal Avenue, maybe a mile and a half south of the original Alliance farm where he came of age as a full-fledged American farm boy in the waning era of the American agrarian Eden.

South Jersey is nothing like the stereotyped industrial cities of North Jersey adjacent to New York. It is rural even today, with fields of sandy-soil, good for growing produce to be trucked to places like Philadelphia and its suburban environs. In its heyday, the small farms that produced vegetables and fruit were the mainstay for many of the settlers and their first- and second-generation children as well as the newer immigrants. When it thrived and built its synagogues and the still existing cemetery where my parents are buried, when those with skills became artisans and those with the dispositions of scholars studied and became professionals, patches of stunted pine trees and scrub oak bordered the fields, and their lives made history. Later, kosher chicken farms supported the new immigrants who came from Germany in the 1930s and those who came after the war from eastern Europe. The back-breaking work, albeit in a free land where people owned their properties, took its toll on the settlers, and as education replaced poverty, a diaspora of this little Jewish community began. Still, vestiges remain.

Chapter 5

Levin/Levinson Family History

Every surviving descendant of my generation and some of the next of the Levin/Levinson family—my two first cousins, numerous second cousins, and their children, any who have any interest in their origins—are familiar with this history. Through the 1950s and well into 1960s, our parents, first cousins and the grandchildren of the original settlers, met regularly for the Levin Cousins' Club. We kids, the second cousins, got to attend the family picnics that constituted the summer meetings which were always in Alliance/Norma. In the late sixties, Beatrice Coltun Harrison, my mother's first cousin, wrote an account of our ancestors' lives. The specifics of the Levin Family history in this chapter and elsewhere are based on her three, legal-sized, type written pages and two later additional pages about her immediate family.

> Our story begins with Baba Leah or Leah Levinson in Kiev, Russia. Great Grandma Leah was married twice; Esther from whom the Levin cousins are descended was the child of the first marriage. . . .

All that is known about Esther's birth father is his family name, Mezeritsky. Esther took the name of her stepfather and half-siblings when Baba Leah married one of two Levinson brothers. Cousin Bea claimed that Leah married Leib Levinson, but birth records from the Alliance Colony and the list of names of the original settlers recognize two couples with the surname of Levinson: Labe and Toba Riva Levinson and Berel and Leah Levinson. Berel and Leah together had three children: Louis, Rose and Bessie, the half siblings of our Baba Esther. How and where the error was made—if indeed it is an error—is lost to the past, but Leah's husband was considered a pious scholar who spent most of his time in study while she supported the family.

Leah, the family breadwinner, owned a little store in Kiev, described as both a *shenk* (a bar) and a candy store. Cousin Bea's quips about selling candy in the bar or whiskey in the candy store shed no light on the nature of Baba Leah's little business, but it could not have been very lucrative. In our era it would be called a "mom and pop" store except that Pop was frail and absent, engrossed in his study of Talmud. The scholarly Levinson died at the age of sixty-two in 1908. Baba Leah lived to the ripe age of ninety, but her heartiness could only endure, not banish the heavy load of sorrow and poverty. She died in 1922.

The resurgence of the pogroms, with the first occurring about 1860, destroyed the *shenk*, but gentile friends sheltered the Levinsons and helped them survive this first disaster. Esther married Yisroel Hirsh (aka Israel Hersh) Levin, a cabinet maker employed by the government, and they enjoyed a good life for a period of years.

> In the spring of 1880, pogroms again seared the community in Kiev as well as other Jewish cities, forcing the Levinsons and Levins to leave Russia. The (cabinetry) shop owned by Esther and Hirsh was completely destroyed, as were most of their personal possessions. The Russian soldiers ripped open feather pillows, poured tar on all the furniture in the shop as

well as in the home, and only some clothing, fur coats, bed *kevant* (feather bedding), brass candlesticks and the samovar were salvaged. Gentiles again befriended our families, hiding them until they could arrange for flight to Germany, their destination.

After much hardship, the families reached Hamburg only to learn that the German Jewish community there was unwilling to assist the Russian Jews who were fleeing from death. Cousin Bea restrains her style when she writes so stoically that the families were "disappointed and greatly distressed" upon being rejected by the good Jews of Hamburg. When I think how disappointed and distressed many contemporary travelers become if their luxury vacations go the least awry, my mind fails to comprehend the anxieties and real-life terrors my ancestors experienced. By this time, Esther and Hersh had six children—Hannie, William, Abe, Fannie, Sam, and a child who died on the ship bound for America.

Through a trio of direct or indirect benefactors—the Alliance Israelite Universelle, the Baron de Hirsch and The Hebrew Emigrant Aid Society—the Russian Jews, unwelcome in Germany, were offered passage to America. The story that comes down to the descendants relates further harrowing details, including the loss of the ship's rudder in a storm which made it necessary to tow the original vessel into the harbor at La Havre, France. Another ship, smaller than the first, was provided so that these families could continue their voyage to America. After forty miserable, seasick days in steerage, made more difficult by caring for five living children, they arrived at Castle Garden in New York Harbor, probably in late 1881 or early 1882. Though they must have been relieved to be on land again, the sorrow of burying a child at sea probably weighed as heavily upon them as the ship's anchor. Bea's account has them arriving at Ellis Island, but the official historical accounts of the Russian immigrants who settled the Alliance Colony state that Castle Garden was the port of entry.

The question persists of whether families were assigned or had a choice of where they would be settled. The story goes that the families arriving in America at this time were given the choice of New Jersey or Oregon. Oregon (in all likelihood, the New Odessa Colony or land nearby) was described as "a wild country" where "eagles carried off children, etc." The "etc." must refer to accounts of livestock succumbing to predators. Whatever the practice, the Levins and Levinsons came to Alliance in May of 1882 to cut-over timberland purchased by the Hebrew Emigrant Aid Society from George Leach and his brother. Families drew lots for small farms that were laid out in small plots and numbered. The families lived in large dormitories and everyone ate at a common commissary, the communal-use buildings being referred to ironically as Castle Garden.

Each family was provided with a two-story dwelling, twelve by fourteen feet, one and one-half stories high, and received a stove, furniture and household goods. During their first winter and the following spring, each family received from eight to twelve dollars for living expenses. The first planting season they also received $100 worth of tools, plants and farm utensils. Family lore relates that each family also received a horse, a cow and some chickens, but these details are not clearly borne out in the historical record. Here facing Henry Avenue, flanked by Max Bakerman's and Eli Stavitsky's houses, Esther and Hersh lived with their five children. Everyone had to work to pay for the farm and the bare necessities of life. Between 1882 and 1888, three more children were born to our grandparents (my great grandparents)—John, Mendel and Jennie.

Severe hardship faced our ancestors. (Great) "Grandfather Levin, for whom so many of the Levin progeny were named, was bedridden for years after a hip injury. Born on November 21, 1843, he died on September 10, 1900, only fifty-six years old. Since there was no picture of Grandfather Hersh Levin taken during his life, the family had him photographed on his death-bed," propped up for what was most likely a posthumous photo.

> The Levin family was impoverished. The older children found work wherever they could, sometimes standing four and five hours in several feet of water to pick cranberries in the nearby bogs. They walked miles to Rosenhayn, Vineland and Elmer, wherever work could be obtained. Then near tragedy struck the family with an epidemic of typhoid fever. William, Sam, John and Fannie took sick but fortunately survived what was then usually a fatal illness. The date was about 1901.

Hannie was the first to marry. She described her wedding in 1891 to Aaron Coltun, son of another original settler family, as very festive and gay. Of their eight children, two daughters and one son survived and were active in the cousins' club. Several second cousins survive in 2018, one active in our current Levin First and Second Cousin Facebook group.

Fannie left Alliance and went to New York City to train for nursing. At some point she married but later divorced and remarried. I have memories of our family standing in my grandfather's driveway to wish Aunts Hannie and Fannie and Uncle Julius a safe trip to Philadelphia and Brooklyn.

Sam went to Philadelphia and worked at difficult jobs to put himself through medical school; he married a teacher, Elizabeth Rudnick, and they had three sons and one daughter. I have no clear memories of Uncle Sam, only stories told by my mother. Elizabeth died too young, unknown to her grandchildren. Their children, my mother's dear first cousins, and their grandchildren were a part of my life, and thanks to social media, I have renewed contact with some of these second cousins. At least two of their granddaughters are named after her.

Abe, the eldest, went to Atlantic City and married Victoria Rosenberg. They established a successful dairy business and had a son and daughter. A grandchild of theirs, of my generation but older, married a non-Jew, and his parents sat Shiva for him, the Jewish mourning practice, because he had married outside of the faith. My own family were not very observant Jews, and I was glad because I thought it so cruel to declare your child dead to you because he or she married a non-Jew. To me as a child, this branch of the family had strict and forbidding demeanors as well as beliefs.

The family had two romances worthy of the tragic love stories of the operas *La Traviata* and *La Boheme* and their tubercular heroines. The younger children, Mendel and Jennie, were in the house with Grandfather during his illness and contracted the disease in their early twenties. Both married spouses from local Jewish families; Mendel married the beautiful Bertha Bayuk and Jennie, Jake Spiegel. The siblings had a single son each but died shortly after their marriages. Mendel lived three days past his twenty-fifth birthday. Aunt Jennie was sent to Denver hoping for a cure, but she returned to Alliance too soon and succumbed three weeks shy of her twenty-fifth birthday. On May 20, 1910, Grandmother Esther died from cancer.

For me, the stars of the Levin family were the sons who lived in the country. William married Lena Bayuk, another daughter of Moses Bayuk, who was one of the two scouts for the initial immigration venture. They remained on the farm in Alliance next to the old synagogue, Tiphereth Israel, and across from the Alliance Cemetery. In 2018, my cousin, Marsha Levin Schumer, told me about Lena's forays into Philadelphia to go the Russian Tea Room for lunch and a reading of her tea leaves during Marsha's childhood, which corresponds to mine.

Their children were I. Harry—named after Grandfather Levin, Jack, Bluma (known as Sis) and Manny, named after his uncle Mendel. Jack married Viola, a beautiful woman with an artistic flair for fashion. In childhood, I adored Viola and their daughter, Annette, two or three years older than I and an occasional playmate. They gave me a treasured gift for an unspecified birthday, a silver and turquoise Mexican bracelet which I either lost or broke, being a rather scatterbrained and somewhat careless kid. I blamed myself and grieved for that special bracelet. Bluma married Dr. Sam Eisenberg who functioned as the family doctor until he specialized. Their son Murray and I have recently exchanged emails sharing some common memories and contemporary interests, but he was older enough in childhood and lightyears beyond me in intellectual achievement at that tender age. Of William and Lena's four children, two—I. Harry and Manny—remained in Alliance or Norma. Their kids, Barry and Marsha, children of Harry and Dorothy (Dot), and Renette (Ronnie), David and Wendy, the kids of Manny and Lil, were my playmates in summers. I loved going to visit Uncle Bill and Aunt Lena's farm, whether with my Pop and brother Mickey and a first cousin or two, or with my parents. I've written about aspects of these visits in poetry and elsewhere in this volume.

∾

Borrowing and Inventing a Grandmother

Because I never knew my grandmother in any role other than that of an invalid and never knew my father's parents at all, I longed for a grandmother. Now in my old age as I write this memoir and remember my Great Aunt Lena and read about Elizabeth Rudnick Levin who, had she lived, would have been another beloved Great Aunt, I can invent and construct a perfect grandmother. It seems that this metaphysical woman I have partially created comes to me in the early hours of the night, when sleep is evasive, as balm for my anxieties and holds me in her arms against her ample bosom, whispering soothing words to ease the physical pain of osteo-arthritis and the sometimes-lingering sorrows over the inevitable losses of life. But because she is a fabrication, a synthesis of parts of other relatives' grandmothers, she remains faceless in those hypnagogic moments when my barely conscious needs summon her.

Elizabeth Rudnick Levin, the grandmother of several second cousins with whom I maintain contact, was married to Sam Levin, the second oldest brother of my grandfather. Three of their grandchildren are writers, so it is not surprising that Elizabeth's prose sparkles as she writes of the "Pioneer Women of the Colonies" in the 1932 *YOVAL*, the booklet commemorating a symposium upon the first fifty years of the Jewish farming colonies of Alliance, Norma and Brotmanville. Long after her death, I—a great niece through marriage who never knew her—admire her language, which is never grandiose nor self-deprecating, but natural and contemporary. She has that observational skill possessed by writers that can endow their own cherished memories with a visible, palpable life of interest to her readers. For her fondness of the early life in Alliance, I have a full, though different understanding of the depths of her emotions and keenness of her memories. If I were writing a novel about four generations of life in an Alliance family, I would make up my own grandmother from qualities of Elizabeth Rudnick Levin and Lena Bayuk Levin, great aunts married to the brash Levin men.

Elizabeth was one of the two first girls from the colony to go to Vineland High School. Aunt Lena and Uncle Bill remained on their farm and were a significant part of my childhood. Aunt Lena's plump, creamy-freckled skin always smelled wonderful, and her affection was wide enough to extend to me. But here I share the writing of Elizabeth Rudnick who, although she spent her early days in the rural community of Alliance, lived her adult life as a Philadelphian.

With the advancing years, recollections of my childhood days in Alliance, New Jersey, seem to become more and more vivid. Incidents occur, however trivial, which evoke reminiscences of country fields and lanes, country boys and girls, country suns and storms, country life.[1]

There is no sentimentality as she pictures the hard-working farm women of her childhood, nor bitter complaint over the work. Her memories of picking strawberries differ from mine in several ways: her experiences with strawberries come from her youth, mine from a spry, poetic old age; she describes the urgency to harvest ripe strawberries and get them to near-by markets because family livelihood depended upon it, while my husband and I, as gardeners, feel an urgency to get them picked, frozen, dehydrated or made into jam for our personal consumption and sharing, rather than sale.

Only the poorest memories can forget the appealing, pathetic picture of strawberry season in our small community! Children trudging along with their mothers, going from one farm to another to crawl along the rows of strawberries, each picker competing with his or her competitor, to see who could fill up more quarts. What mattered blazing sun, or parched throats, or aching backs when one understood that these precious berries had a short life, and that near-by markets greedily devoured the farmers' entire production![2]

I wonder what she would make of my poem/meditation about John Lennon and picking strawberries. I think if we could meet in a space-time warp for lemonade and my grapefruit scones with strawberries and cream, she'd get it and laugh, as if she had always known who John Lennon was.

A Bone of Contention with the Ghost of John Lennon
Over Strawberry Fields Forever

last year i turned each unripe berry's curve to the sun,
creeping through a single row; their red hearts and green leaves
dotted by starry blooms. they were few enough.
there was time for such indulgent care and contemplation.

this year with mother plants settled in rich soil,
the unruly daughters run amok by the dozens,
march-dancing fifty feet in and out of a quadrupule-wide row,
their leaves in mudras of ecstatic hand jive,
declaring their wild passions.

i crawl on hands and knees, between berries and beans,
muttering under my breath, in contention with the ghost
of John Lennon and his audacious dream of Strawberry Fields Forever.
his romantic idyll has become my spring curse.

i like his music well enough and am always willing to give
Peace a Chance, but the concept of infinite strawberries
is appalling. forever is married to everywhere.
eternity walks hand-in-hand with every arable spot of land.

in my garden those amorous red-hearted girls of spring
spread their juicy sweetness to the cucurbita, threatening
to swallow squash, pumpkins, melons, and cucumbers.
we are already sated, our freezer filling, jars of jam stacked high.
did John Lennon ever fill more than a priceless Japanese bowl
with ripe berries? did he arrange five of them gracefully on an
antique satsuma plate Yoko probably scarred with a diamond drill
to make an artistic statement while lying nude together in the fields?

did he ever ache in every joint from picking gallons a day?
did he wonder how to save them for winter and buy a food

dehydrator, stock up on canning jars and baggies, consult
cookbooks for shortcake and pie recipes with a twist?

oh, John Lennon, i'm sorry you are dead. if you lived still
i would invite you to my garden and let you pick from one
row of the ruby jewels, the red berries of passion, and gorge
you and your love with sweet flavor for an afternoon.

∾

But for me, the polestar of the Levin firmament was my
grandfather John H., the first child born in America to his
family on February 22, 1883. After marrying my grandmother, Bessie Barish, and having at least my mother and my
middle aunt, they moved to Atlantic City. My parents met
in Atlantic City, married, and later moved to Philadelphia.
At some time in my first few years, my grandmother had a
severe stroke and thereafter resided in a nursing home until
her death when I was eight.

Pop had moved back to what I call South Norma, probably in the mid to late 1930s. I no longer remember whether
Nanny Bessie returned to that section of the Alliance Community where she settled with her family as a teenager, part
of the group of Russian immigrants who followed the original settlers. By the time I was enrolled in kindergarten, the
three sisters (Sylvia, Jeannette or Jean, and Esther) and their
families all lived within blocks of each other in the Oxford
Circle neighborhood of northeast Philadelphia. Toward the
end of my grandfather's life they took care of him between
his hospital and convalescent home sojourns. We lived the
model for a close extended family, warts and all.

The implication in the brief Levin history related by cousin
Bea Harrison is that my grandfather and his brothers, while
growing up on Henry Avenue in the heart of the rural community of Alliance, were a rowdy lot. Bea Coltun Harrison

tells us in her family history that "while there is no record of a great scandal in the Levin family, the Levin boys had the reputation of being rough and tough, and the story goes that they were responsible for more fights than anyone else and spoiled many a country dance by rough housing that ended up in a free for all."

In my youth and young adulthood, I knew that my grandfather had frequented dances in his younger days because he loved to waltz. At every family wedding and bar mitzvah, Poppy and I waltzed together. When I first read Beatrice's account of the Levins in 1966 or 67, I was shocked to learn about the propensity of the boys to rumble—a term from the 60s that seemed to me best applied to urban gangs and romanticized by *West Side Story* but the term that came most readily to mind. It was inconceivable at the time to put my grandfather and his brothers in that context. My cousin Liz Kelner Pozen, a granddaughter of Dr. Sam and Elizabeth Rudnick Levin, tells me now that they were brawlers and fighters and that rumors existed that they had been horse thieves in Russia—unfounded rumors, and quite possibly one of the old stereotypes regularly employed to slander Jews.

As a child I was socialized to be squeamish and fussy; that was, after all, the era of "little girls made of sugar and spice and everything nice" and "little boys" made of "frogs and snails and puppy dog tails." The repulsiveness of contemplating detached puppy dog tails and gender stereotyping aside, I had a deep, innate desire to get over my fear of snakes, my disgust of rot and slime—all the scary and icky things overly-protected children scorned. My somatic type—average height but thin, even downright skinny as a kid—inclined me towards an active child's lifestyle. In fact, I was a combination of a finicky girl who loved pretty dresses, ballet tutus, and Halloween costumes and a semi-cautious tomboy who sometimes overestimated my own abilities. Like a cat, I often impulsively climbed high into the mulberry tree and then called for help in descending once I looked at the ground which seemed such a remote distance from my arboreal perch. All children with access to the natural world have a healthy curiosity and fascination with what they discover; they are all willing to take risks and get hurt until their overly protective parents hurl a battalion of warnings at them.

I loved my grandfather and his home in the country because of the portals it opened into the natural world and of how deep-in-the-bone it was bred in me and what I would yet become. Our entire family always associated the song "Nature Boy," recorded in 1947 by the Nat King Cole Trio, with Pop. However, when I listen to the lyrics, I realize how unlike Pop the song's subject—based in part upon the life of the writer, Eden Ahbaz—really was. Pop was neither shy nor philosophical. He never wandered far from the location of his birth, and although his parents traveled far "over land and sea," he made no trans-oceanic voyages; indeed, he probably never went farther than Philadelphia, Atlantic City, Maryland, and possibly as far away as New York. Attributing the qualities lauded by the song's lyrics was just another layer of material applied in the creation of the family myth of our own American Pan, our first-generation farm boy, our faun of the baseball diamond, our minor god of the natural world.

Any pretensions I am guilty of holding in my life have to do with my longing to have been from a more educated, refined, cultured family than mine. Even without one's treasured pretensions, it is not easy to own that your beloved grandfather was a rowdy boy/young man whose behavior you would not have condoned in your youth nor at any time in your life. It's only legend now but believable. He was a country boy with boundless energy, four brothers and neighbor boys with whom he had grown up under always demanding conditions. Our parents were of a different generation and their parents and grandparents further and further removed from our current realities, the issues and perspectives from the second half of the twentieth century to 2019. I am wary of romanticizing my experiences of childhood, as much as I am skeptical of people who romanticize my life of today: the seemingly simple life of a semi-subsistence farmer/ artist-gardener. I hold out my hand to my beloved, imperfect grandfather, inviting him, in this present time, to come down from the pedestal I had built for him.

Chapter 6

The Home of John H. Levin, "My Redneck Zeyde"

I call my Pop my "redneck zeyde," a loving, humorous term that requires defining for the non-Jewish reader who has not had close relationships with Jewish culture. "Zeyde" simply means grandfather in Yiddish, and most Jews of eastern European backgrounds around my age referred to their grandfathers as "zeyde" and their grandmothers as "bubbe." Pop rejected the language of the old world for an American title.

The first eighteen years of my life were anchored by significant periods in the summers and frequent weekends at my grandfather John H. Levin's farmette where he gardened, raised fancy poultry and dairy goats at various intervals, and ran an egg route to Philadelphia. My mother, Sylvia Weinstein, and her two younger sisters, Jean Bailine and Esther Ernst, took turns in the summers well into the 1950s to be there with their children for the sake of a close, nuclear family that knew its roots. I have no memories of my grandmother, Bessie Barish Levin, living at the house. She came to the South Jersey settlement as a young teenager with her parents in the next wave of emigration from Russia and never possessed what has always been, for me, the cachet of being an original settler. She died in a nursing home when I was eight years old after years as an invalid. The stroke she had sustained left her paralyzed but able to say my name haltingly for a while when we visited her in my earliest years. For all my life, what has persisted is the visual memory of her shorn, white hair, like a paint-dried brush, against her pillow. I do remember seeing among my mother's photos a few pictures of Nanny Bessie. She was cute in her youth—short, dark and plump, with a shy, sweet smile. My mother, Sylvia, aka Cookie, for how cute and fun she was when young, took after her mother.

John Levin's home in rural South Jersey was the place where I learned the names of wildflowers, first felt the warm flank of a milk goat against my cheek, and sensed, with both excitement and fear, the wild world beyond the sidewalks of the city. It was where the lure of the natural world declared itself a magical and mysterious realm, a world which had claimed me as one of its devotees though it took many years for me to return to a country setting as my dwelling place. It was the place where I learned the value of cultivating a patch of wilderness in order to grow food with my own hands and of caring for animals that would provide me with milk and eggs. It declared that even with tractors and tools, even with dedication, muscle and commitment, the effort of a human to shape a wild environment, the cultivation of a piece of land surrounded by wilderness, was a tentative and temporary endeavor.

In my childhood summers there, I was a little country girl who gently "tied" earthworms in "knots" to see them wriggle free and then, all her life afterwards, rescued earthworms when rainstorms washed them into danger zones. That little girl has become an old woman who forgets to put her gloves on while gardening and does not give a damn how old, wrinkled and stained her hands are but who, nevertheless, has an urbane and sophisticated aspect. I like to "cleanup good," don my best duds, and go to the city for opera and chamber music—indeed many kinds of music—and ballet and modern dance performances. I love theater and foreign films, art museums and galleries, poetry readings and slams, good restaurants and food trucks—all primarily urban by default—but find that a little goes a long way. I have always been easily overstimulated and need more solitude and nature than people and culture. At one point in my life I thought that maybe a city home and life and a country get away would be mine. That never happened; I'm not good at making money, and a simple, earthy country life has become the one and only life I live. I'm my Pop's granddaughter with my own identity.

My Poppy's older brother William, my great uncle, and his wife maintained an even more substantial farm where Aunt Lena made sour cream from their cow's milk. In my poem "The Legendary Tomatoes of New Jersey," I employ license by adding butter and cottage cheese to her repertoire. In their farmyard, huge crows flew around the corn crib and pecked the kernels that fell to the ground as the slats of the crib made patterns of light and dark bands against the setting sun.

Across from the farmhouse that was once theirs is the Alliance Cemetery. My grandparents, my parents, my aunts and one uncle, and my cousin Bruce were buried in the small Jewish cemetery across from where William and Lena Levin's farmhouse still stands. The practice of traditional Jewish burial for my family ends there with them. My brother's body was cremated, and his ashes were scattered by his wife and children in the waters off Hilton Head Island, where he loved to vacation. My body, also, will be cremated and my ashes dug into a spot near our garden with a sour cherry tree or blueberry bush planted in the ground, except for a spoonful of ashes that will be scattered in the Buffalo National River. Perhaps, my husband will send a spoonful to my cousin in Alliance to scatter in the Maurice River, thus marrying the tea-brown water of the river of my childhood and the minerally blue-green of the river of my Ozark life through a pinch of my earthly remains.

❧

Forty-three Long Miles or "Are We There Yet?"

My earliest memories of the drive from northeast Philadelphia to my grandfather's bungalow on Gershal Avenue, south of the railroad station in Norma, New Jersey, include a bout with car sickness on nearly every trip. What, to my childhood self, seemed an interminable trip could not have been much better for my poor parents whose hopes of making the drive without having to pull over and care for a puking child were once again dashed when my plaintive cry of "Stop the car!" shattered the calm of drive. It took years for me to outgrow my motion sickness; even in junior high, the movement of the tour bus and my state of mental excitement about going on field trips could still erupt without warning into a siege of vomiting. Another bead on the string of events responsible for childhood humiliation and woe that, fortunately, most awkward children outgrow.

On summer's holiday weekends, bridge traffic from Philadelphia—whether the Tacony-Palmyra Bridge, connecting the Tacony neighborhood of northeast Philly to a village northwest of Camden, or the Benjamin Franklin Bridge from downtown Philadelphia into Camden—was an agonizing snarl of cars brought to a total standstill. In those days, cars were not air-conditioned, and the air was thick with the smell of exhaust and a mirage of shimmering, palpable heat. I usually fell asleep on the return trips and was carried into the house by my daddy and put to bed.

However, on the way to "the country," on the still cool morning drives to my Poppy's place, if I escaped being sick, or when I had recovered from my "spell," anticipation inflated me as helium inflates balloons. I felt that I could almost exit the car and float there had I tried hard enough. Passing through Haddonfield's colonial neighborhood, we always acknowledged the house with the sign proclaiming that it had been a stop on the Underground Railroad. The Philadelphia School System, during the golden age of public education, provided us with good, if not complete instruction and inculcation. My school days (1945–1959) fell well into that period, and I was pleased as a child that we lived in Philadelphia, the cradle of liberty, and were familiar with its environs and Quaker history.

But once we passed the Camden-Cherry Hill area, the almost two-hour drive along two-lane New Jersey Route 47 took us slowly through what were then the small towns of Woodbury, Glassboro, Franklinville, Malaga and then led directly to Gershal Avenue, through Brotmanville and into the heart of Alliance, then Norma. Most of the drive linking the small towns was rural, and then the sight of the synagogue in Brotmanville signaled to me that we were THERE, we were in "the country," the small, rural community where we somehow belonged. Whenever we had friends along who had not previously been to the country with us, I proudly pointed out the homes of my mother's relatives: her first cousin, I. Harry Levin, Esq., who was named not for his more famous Bayuk progenitor but for Israel Hersh Levin, my great grandfather (and his wife Dorothy, known as Dot, and their children, my cousins Barry and Marsha); then his parents' farm, my Great Uncle Bill and Aunt Lena. Across the road from Uncle Bill and Aunt Lena was and still is the Alliance Cemetery where all our dead relatives were buried. I pointed out the cemetery as the final resting place of our dead relatives. Farther down Gershal, bordering the old Alliance Community and Norma, lived my mother's cousin Manny Levin who made his living as a chicken farmer. Manny and Lil's home, like ours, was a place where there was always enough to feed any and all relatives who popped in to say hello. Their children—Ronnie, David and Wendy—were my playmates and are now Facebook Friends.

Finally, we came to the junction of Gershal and Landis Avenues. Landis Avenue is the wide, boulevard-like street that runs the length of the city of Vineland. I remember the feed mill owned by the Berkowitz Brothers and think that I remember it first on the northeast corner of Gershal and Landis, but I have recently been corrected. In the late 1940s or early 50s, there was a fire that either burned the mill to the ground or damaged it beyond rebuilding on site; the mill was relocated after that. Every rural location in which I have since lived naturally has a mill, and the dusty, grainy smell of these more recent mills—Quakertown, Pennsylvania, or the various mills in Arkansas where we shop—takes me back to the Berkowitz mill and my early childhood summers. On the northeast corner stood Joe Gaidosh's general store. At that junction of Gershal Avenue the concentration of the Jewish community was not quite as dense. Closer to "downtown" Norma/Alliance, the Brenners and a family, possibly named Reisman, owned smaller grocery stores, but at Pop's end, Gaidosh's was it. Joe Gaidosh and his family were not Jewish, probably the half of one percent in that enclosed community who were not. We were so proud once we were old enough to walk to the corner and make purchases on our own at Joe's. Across from Joe's was Fox's Garage. Izzy Fox and his family were Jewish. He and his son Ellis, who, I am told at the time I am writing this, is still alive, were amazing mechanics. However, the mixed smell of motor oil, gasoline and dirt and the sight of men with grease-smeared hands and faces were overwhelming to a fussy, car sick-prone little girl. Even the combustible engine, transmission grease, brake-fluid and hot metal smells in a garage, without the motion of a moving vehicle, made me feel ill. My grandfather, my father and uncles, and every male of driving age residing or passing through the area who ever got behind the wheel of a vehicle, depended upon the automotive and mechanical genius possessed by Izzy and Ellis Fox.

Just past the Foxes' garage were the railroad tracks, and then we passed the houses, mostly on the right side of the road, that belonged to friends and family. To you reader, who like a friend visiting the country with us for the first time, I say that we are so close now that I will take you straight there; later there will be time to visit some cousins and neighbors, Pop's community up and down Gershal Avenue.

∾

The Bungalow on Gershal Avenue

But here, we have arrived at my grandfather's. Finally! Are you, like the young girl irritated by the ride, impatient with the minutia of my precious memories, and as ready as I to explode from the confines of the car into the country paradise of John Levin's five-acre farmette and comfy bungalow? We pulled into the gravel, horseshoe-shaped driveway of Pop's on the south side and parked in the back. Only strangers and official callers knocked at the front door; everyone else, like all country folk, knew to come to the back door and enter the heart of the house, the large country kitchen.

Here, in my enthusiasm to explode into that world, I am scattered, wanting to show you everything at once. I take your hand and pull you into the yard where the hammock hung from a post and tree, either the black walnut tree or the mulberry tree which dropped its purple jewels in August. That I can no longer remember which tree held the hammock feels like a larger loss than it probably is, but I am distressed by my own acceptance that another memory is gone. Any ability to recall a third tree in the trio is gone as well: maple or oak? Or could it have been an evergreen? Like a precious stone from a necklace of beads, each one meaningful, a memory has rolled away and is lost to the earth, invisible in leaf litter, irretrievable from the leaves of grass. A consolation memory

says there were no conifers on that side, only strong, decid-
uous trees. The mulberries—whatever the precise location
of that drupe-bearing tree in the yard—we gathered and ate
until our hands, faces and bare feet were stained with their
delicious juices.

The primary feature of my grandfather's place was the
close to one hundred trees he planted, over a period of about
twenty-five years, on what was probably no more than a
one-acre piece of his property. The larger trees cast shade for
the kids who fought over and were forced to share the one
stout, canvas hammock, and for the adults in lawn chairs.
The smaller trees, like the grandchildren, grew in front of
everyone's eyes. He planted maples along both edges of the
driveway and its arc in the back of the house where the drive
separated the house from the garage and pens of the goats
and poultry: silver maples, with shaking, shimmering leaves,
Norway and sugar maples, too. Young weeping willows grew
in the area we called the "cesspool." I wonder now whether he
had a septic tank and drain field and question the sanitation
of the waste drainage area, but in my childhood I imagined
it a forbidding moat and leapt easily across its narrow span
to enter the privileged royal environs.

In the front of the house, he planted a tulip poplar,
a maple, a blue spruce in the southeast corner and some
junipers or cedars. In 1995, my brother and sister-in-law
and I, returning to their home in the Maryland suburbs of
Washington, D.C., from a reunion at the Jersey shore of our
father's dwindling family, took a detour to stop at the Alliance
Cemetery where our parents and grandparents are buried
and to drive past our grandfather's place. Nothing looked
familiar; the asphalt business that had bought his property in
the early 1960s had changed so much, including the house.
Although they left many of the planted trees on the lawn and
much of the wooded area, it all looked different, and we felt
profound sorrow and loss. Then I cried out, "Mickey, look!"

and waved my hand wildly at that blue spruce tree, the only identifiable remnant of Pop's arboreal labors. Each of those trees was irrigated with grey water from the ringer-washing machine that Pop emptied by the bucketful along the dripline to encourage root growth. Long before EPA regulations on phosphates in laundry detergents, high levels of phosphorous began accumulating and were responsible for eutrophication of our streams, rivers, lakes and even parts of our oceans from the late 1950s on, but Pop's trees grew with those phosphorous boosts. Via a Google search, I learned that the first National Arbor Day was in 1872 and that the National Arbor Day Foundation was established one hundred years later in 1972. Although the first was well before Pop was born and the second well after his death, I cannot help but think that some organization or agency must have existed to encourage the greening of America through the planting of trees and that Pop happily tapped into whatever structure there was to get those hundred trees which he planted in the yard.

Depending upon the season and years, we could continue to explore outdoors. If it was deep summer, we would go see the small garden where my Pop grew a small strawberry patch, an asparagus bed, a few tomato plants, cucumbers, herbs and garlic, watermelon and cantaloupes, and sunflowers. Corn and green beans were usually ours for the picking because he rented the majority of the five acres to local truck farmers who grew their crops in the fields that separated us from the neighbors on the north and from the mysterious allure of the woods to the south and west. We also had the option of shopping for fresh area produce at the Kotok Store and Farm Stand on Landis Avenue between Vineland and Gershal Avenue.

Behind the house and on the other side of the driveway was a spacious garage that housed Pop's vehicle. I only remember panel trucks with throttle, clutch and stick shift on the floor. The hand brake was a huge lever which required more strength than my skinny, little arm could muster in order to safely park the vehicle. Every country kid I have ever known learns to drive early, permitted by parents or grandparents to first steer the family car or pickup truck along lightly trafficked roads, whether paved or dirt. My brother, cousins and I were granted the same privilege once we were old enough. I never progressed too far in my younger days long before the era of power steering because, even with both hands driving down the flat, straight country road, steering seemed so hard. I managed well enough with help, but comparing my struggle to the elegant nonchalance with which Pop kept the vehicle moving straight ahead, using only one hand on the Bakelite knob attached to the steering wheel, I felt that driving was well beyond my skill level.

The garage was filled with chains for snowy and icy weather, extra tires, gas cans, jumper cables, jacks, a tire pump, and materials for patching inner tubes, wrenches, a whole host of tools for outside work, the push lawnmowers, and gardening equipment to till, hoe, weed and harvest what he grew: everything necessary and all in perfect working order, clean and in place. The huge tractor innertube we took to the river hung on the garage wall. On one of the wall studs in the garage rested the oilcan, used for lubrication of various mechanical parts of machines and tools. I loved the oil can, no larger than the closed fist of a big man, with its jointed spout, out of which spurted the viscous substance that banished squeaks from, and made smooth, the moving parts of things, and freed them of rust. I loved holding it in both of my small hands and pushing the bottom of the can to release the tears of oil.

At that time in my life I had a wonderful children's book about personal hygiene. Set in a gleaming tile bathroom, the illustrations brought to life the tube of toothpaste, the toothbrush, the cake of hand soap, all animated with movement and emotion. The personification of the items in the book was so significant to me that it transferred to many of my grandfather's tools and the accoutrement of his garage and workshop. Somehow, that oilcan seemed to possess an awareness of the child who found it so intriguing. A can of 3-IN-ONE HOUSEHOLD OIL, an essential item for house and farm, is always handy for me or my husband to use. It serves the same purpose my grandfather's oilcan served but completely lacks the charm that simple implement held for me in my childhood.

Also, in the garage or one of the smaller sheds was the wooden wheelbarrow with removable sides that transported not only heavy, bulky objects like buckets of manure and sapling trees, but also me for rides around the property. I so loved the flat bed of the wheelbarrow and the sides that could be taken off and put back that I endured the bumping against my protruding vertebrae as my brother or Poppy wheeled me around. My brother Mickey and I even played wheelbarrow on the grassy areas of the yard where Mickey would hold my feet as if they were the handles of the barrow and I would walk up and back on my hands.

In the same general area as the garage were the pens, hutches and coops for the goats, rabbits, and poultry that were part of Pop's experiment with five acres and independence over the years. This phrase, from the title of the book *Five Acres and Independence: A Handbook for Small Farm Management*, by M. G. Kains, republished in 1973 by Dover Publications, had great significance in the back-to-the-land movement of the 1970s, although the original first edition was published in 1935 by Greenberg Publishers of New York. When I asked my husband if he remembered it from our early days together, when we were working and planning to buy our land, he was vague about recognizing the title but certain that it was never in our homesteading library. It occurred to me then that my first awareness of the small book must have stemmed from seeing it at my Pop's when I was a kid. The ethos and information held in its pages were what constituted his farming life during my time spent in the country.

South of the structures that sheltered the animals was a cracked concrete pad not much larger than ten feet by ten feet. Whether an outbuilding had ever existed there during Pop's time on the property, I don't know. In the cracks of its surface grew weeds, but even the weeds there were marvelous to the inquisitive mind of a young child. Sometimes we would find tiny wild strawberries, no larger than the tip of my little finger but of such concentrated sweetness that a few berries provided a delectable treat whenever we were lucky enough to find them. The ferns of asparagus grew tall, feathered and green among the cracks, having sprouted from the fruit of the female asparagus that birds scattered in the fall. The weed redleg lady's-thumb (*Polygonum persicaria*) grew there. Its tiny pinkish-white flowers, like clusters of seed beads on notched stems, seemed beautiful and delicate as a precious necklace to me. I picked them for bouquets for my always appreciative mother. About fifteen years ago this same weed, now as unwelcome an invader in the garden as henbit and deadnettle, began encroaching upon my flower beds. I realized that I had imported it in the soil of potted plants received from a dear friend and gardener extraordinaire who lives in the charming artists' colony of Victorian gingerbread houses in Eureka Springs, Arkansas. Fortunately, I had long since accepted that the gardener is always at the mercy of wildness and contented myself in reuniting with this once beloved "flower" of my childhood. Nevertheless, I still pull every branching, knotted stem and root that I can yank.

Beyond the lawn and buildings and the garden and livestock areas were the fields rented to local truck farmers, and then beyond the fields were the woods that led to terra incognita. When very young, I never ventured beyond the boundaries of the homestead alone, except two or three feet into the cornfield, when even as a two-year old I would pull down my white cotton panties to pee in the field rather than return to the house to go to the bathroom. Pop would sometimes lead us into the woods for nature walks, but occasionally, when my brother and cousins felt defiant and daring, they would march through the fields and into the edges of the woods. Summoning my courage, I pumped my little legs to keep up with them. Southern New Jersey's sandy soil is like a sponge and holds boggy areas where various wetland wildflowers grew. I have a strong memory of sweetflag or calamus (*Acorus calamus*) from the boggy woods beyond the fields of Poppy's five acres. So many years later while hiking near our Ozark home, my husband and I came upon an area of land wet with run off from spring-fed ponds, and even before plucking and rubbing a piece of the blade-like leaves, I could smell the distinctive aroma. This was sweetflag, the same calamus I knew from childhood wanderings in the woods near Pop's house. Only a few miles away were the cranberry bogs where my grandfather and his siblings—and possibly even his mother, my great grandmother Esther—walked and stood for hours picking berries to help support the large and struggling family.

Also, in the woods were snakes and ticks. The snakes we saw slithered away through the layers of leaf mulch. Pop caught snakes and kept them in jars so that we could learn to identify them and then set them free far from the house and yard. The ticks we brought back already embedded in our flesh by the time we reached the house. My mother or one of the aunts, brandishing tweezers and reiterating warnings about exploring in the woods, grabbed us and dug into our scalps or tender flesh to extricate the dreaded pests and then drenched the spot with Dickinson's Witch Hazel, iodine or mercurochrome to disinfect the bites. That I ever moved to the Ozarks and have stayed living in the middle of forty wooded acres after being de-ticked by a zealous mother some might consider either mad or miraculous.

As the sun reached its zenith and beat down on us, the house and its treasures, which were variable and delightful to the senses, beckoned. The entrance via the back door, with its terrifying (in my earliest days) flight of steps without backs, led to Pop's lair—the large, tidy basement. The coal furnace occupied the center and an array of my grandfather's tools declared this space the male-centered world where things that made the house function and things from the past were stored. From the landing that led to the basement a few steps led to the kitchen door. The iron skeleton key that locked the back door and other keys, flashlights and batteries, jackets, hats and boots hung there or sat on a couple of shelves. That space, scarcely large enough for two standing adults to occupy simultaneously, could have been broken into so easily. In the mid-50s when my world expanded to include a teenaged social scene, it became, on a few frustrating and embarrassing occasions, a Houdini entrapment, when that skeleton key would jam with me inside wanting out or outside wanting in, but when anyone was at home during the day, the doors were unlocked. It took deft fingers to turn the doorknob exactly so and open the door to the kitchen. It always seemed like I was waiting in the wings of a stage, ready to make my entrance into a theater of another world, the perfect setting of a country kitchen.

In the heat of summer, we first rushed to the large enamel sink and turned on the tap for a delicious drink of cold, artesian well water. On every return trip to the city and its overly chlorinated water, we carried home as many bottles of that water as we could fill and fit into the car. We drank deeply and slowly, refilling our glasses to quench our thirst after the long drive from the city. Once hydrated and settled, I turned to my favorite thing in Poppy's kitchen, the queen-of-all dry kitchen cabinets, which occupied most of the west wall and held fascinating, magical-seeming kitchen tools. The cabinet, a deluxe model even then, probably became a prized possession of a 1970s back-to-the-land hippie woman—or anyone with a country kitchen look in her or his house—coming across it in a chic antique and collectible shop anywhere between Cape May and Philadelphia. Known as dry kitchens because of their tin bins and hoppers for flour and sugar—or as pie safes in the south—I bought one in the late 70s. First, my husband and I kept it in our original rough-cut kitchen, then moved it into our attached, passive solar greenhouse after our kitchen cabinets were built. Now in the last part of our lives, it is still here, with a fresh coat of zucchini-green paint, on our screened porch. Our electric grain grinder, which my husband uses to freshly grind wheat, rye, and spelt berries, and buckwheat or oat groats for his sourdough bread, has its permanent spot on the recessed shelf, allowing us to use the chipped but still functioning, pull-out enameled counter for food and the accoutrements of summer living on the porch.

Pop's was white and mint in condition. It had a flour bin with sifter at one end and a sugar bin at the other. Massive and majestic to the small child I was, it lingers in my memory, its double doors filled with essential canned goods, baking supplies and utensils between the flour and sugar bins. Not only did it have the pull-out, enameled countertop, but it also possessed two other more complex mechanical wonders that always worked smoothly. It had a roll-top hood to close off the recessed section behind the counter where the country wife or daughter could store her cut-up fruit and other ingredients while rolling out her pie crust on the counter. My mother and aunts were all accomplished home bakers, each with her own specialty. I can picture my mother in her cotton house dress and apron, crimping crusts to pre-bake the shells for her chocolate or lemon cream pie or rolling tight the whipped cream and strawberry-filled "jellyroll" cake.

I was restricted to how often I was permitted to pull this slatted feature of the dry cabinet up and down, and when very small, I was too short to reach it. The bottom doors exhibited the true genius of simple mechanical design and excellent hardware. When opening both at the same time, a special hinge enabled the opening doors to pull out the main shelf inside. I believe that a row of graduated drawers flanked these wonders of kitchen engineering of that era. I spent hours in simple play with the items found in the drawers and doors, whether the old-fashioned flour sifter or the collection of glass globes from coffee percolators that had long since been discarded for a host of reasons. I imagined them as minute crystal balls that could predict the future and dressed up as a fortune teller to lend a dramatic flair to my daydreaming. I must have entertained myself from age five to ten in front of that dry kitchen as I grew older and capable enough to help with cooking and baking and to attempt my own ventures at solo food preparation.

When my grandfather's health began the long decline between 1959 and 1960–61, and the property was sold sometime after, I grieved for the loss of that piece of wonderful furniture in a way that I could not explain. At age eighteen and nineteen, I could not have read in the little array of crystal balls my future of sixteen years hence when I would move to Arkansas.

I often think that my visual cortex is the most active part of my brain. As I write about my grandfather's dry kitchen from the comfort of my own kitchen, warmed by a fire in the woodstove, I can visualize that treasured piece of furniture filled with mason jars—half pints, pints and quarts—packed with red and garnet-colored jams and jellies, the golden greens of pickles and tomatillo salsas, the quarts of applesauce, canned tomatoes, and vegetable soup. To have that cabinet packed with such a wealth of one's own labors is a symbol of deep security to me.

A tall narrow, metal cabinet that held canned and packaged goods stood between the "Queen of Cupboards" and the door to the mudroom. We loved putting decals on it, and I remember decorating it with swans and bubble-blowing fish. Smoothing the film of wet decal onto the surface was challenging, and I judged myself harshly if I could not completely smooth out the wrinkles. The imperfections of the decal's surface, however, left open an opportunity to scratch, scrape and peel off the defective cellophane decoration and start again. They were not the plastic-coated paper that peels neatly from a non-stick backing like bumper stickers of today, but a kind of cellophane that required soaking for a minute in warm water so that the decorative image could slide from its backing in one slick swipe onto the surface one wished to decorate. Far more challenging than today's version to apply but more sustainable, I suspect. The decals may have come in boxes of cereal, so my mother was pleased when I agreed to eat more oatmeal, or perhaps they came in boxes of Cracker Jacks, the sticky, sweetened popcorn and peanut treats.

On the short, north wall of the kitchen lived the accursed propane refrigerator during the 1940s until it was replaced with an electric model at some time in the early 50s. To its left stood a table or hand-built shelving unit which held the large ceramic crocks in which Pop's kosher dill pickles, both cucumber and green-tomato, were brined and fermented in summer. My mouth begins to water with this memory, and as the movie stars in some preposterously-themed musical burst into song and dance, I, the writer, seize the license to include the appropriate poem to further grace that heart of the country home, the large, living kitchen.

❧

The Legendary Tomatoes of New Jersey

The legendary tomatoes of New Jersey linger
on the taste buds of my childhood memories,
their tart-sweet juice dripping warmly down
my chin and arms, right in my grandfather's garden.

That South Jersey sand could really grow
the best asparagus, sweet corn and tomatoes
that anyone had an appetite for in the springs
and summers of my youth. True, they were
not grown organically, it being the 40s and 50s
when DDT was king.

And I don't remember the varieties of tomatoes,
those little plum ones that someone's Italian
grandmother turned into rich gravy and
the big hunking slicers we ate with charred,
high-quality burgers and grilled corn.

Green ones and cukes made it into the crock
with dill and garlic from my grandfather's patch,
and once they were fermented just right went
into jars in the old propane fridge. The grated
black radishes were eaten by the brave
with rye or pumpernickel and Aunt Lena's
butter, sour cream and cottage cheese.

And my mother and her friend: the two Sylvias,
dark beauties with red lipstick and cigarettes,
canning peaches and tomatoes in the other Syl's kitchen.
Later, laughing in their orangey-red sundresses,
two hot-tomato young mamas, saviors of summer's bounty.

∾

Next to the continually troublesome, Servel-brand propane refrigerator and the later electric version was the small hallway that led to the bathroom and two bedrooms. We vied for space to stand on the grate when visiting in winter for the warm blasts from the coal-burning furnace directly beneath us in the cellar. We shall explore that part of the house later.

The kitchen table sat against the interior wall that separated Pop's bedroom and the living room from the kitchen. It must have hailed from the late 30s, prior to Formica tops and vinyl-covered chairs, because the legs and sides were made of wood and the top, enameled and stenciled with a simple design of lines and corner details. A drawer for the flatware faced outward, and all through my childhood I had a fondness for setting the table, opening and closing that drawer that always seemed to stick. The chairs must have been wood, but I have no distinct memory of their appearance. However, the slightly gluey or sticky feel of the aged finish on their wooden backs and legs is a tactile memory that almost makes my fingertips tingle with the impulse to scratch the wooden surfaces. When just one sister was present, my mother or either of her younger sisters, with her own child or children and a cousin as well, we kids were often fed first. We ate simple salads, sandwiches and vegetables for weekday meals at that kitchen table.

All of us kids vied for the privilege of making toast at breakfast. The old-fashioned toaster was a flat-bed model with a space large enough for two or possibly four slices of bread at a time. It consisted of electric coils protected by a wire grate. A hinged grate held the bread, much like a fish basket that holds fish securely upon an outdoor grill, and was turned to toast the second side when the first was done. If the person in charge of making the toast was attentive, perfect toast was the reward. The woven cover on the electric cord probably began to fray or the toaster ceased functioning, and it was eventually retired and possibly replaced with a pop-up model. I can still smell the hot electric coil and heated bread.

Generally, we were well-behaved kids but sometimes something would strike us as hilarious and we would jack up the humor until we laughed uncontrollably, and milk or the "bug juice" we mixed from orange, pineapple and grape juice streamed from our nostrils or we coughed up half-chewed food. The more a mother/aunt urged us to calm down and behave, the more uproarious we became over our little jokes. One particularly notorious incident occurred when my cousin called my mother "Aunt Silly" instead of "Sylvia." In probable desperation for a moment of peace and quiet, she sent us outside, a reward rather than a punishment for our wild mirth. We loved Kool-Aid but were limited in our consumption of that and soda as well as the hard candies Pop kept in a tin or the occasional boxed chocolates that made their way into the house. Sometimes in summer we were treated to glasses of ice water with a spoonful of strawberry or other fruit jam stirred into the cool, delicious well water. On the weekends, when the father/husband of the family in residence came from Philadelphia, meals were more elaborate and included picnics on the lawn with neighborhood friends, or friends and relatives from the city.

The doorway to the living room was next to the table. Above the door hung Pop's rifle on a rack made from two front deer legs bent at the knees. I loved to stroke the hair on the legs, but having had my childhood shaped by Walt Disney's "Bambi," I grieved over the loss of the living animal. The gun, probably a .22 caliber rifle, I had no interest in until I was an adolescent or teenager and Pop taught me how to shoot it. As much as I wanted to learn—a privilege the boys secured before I did—the recoil of the butt into my shoulder and loud report were off putting for me. I still feel the same way about guns, noisy, dangerous but sometimes necessary tools of country living; however, I have not practiced shooting our .22 in years.

The short kitchen wall next to the doorway was taken up by a yellowed pine cabinet, probably constructed by Pop himself. The top section had glass-doors revealing Depression Glass cups, saucers and dessert plates in green and pink; the regular, every-day dishes; and a brown-glazed, incised earthenware jug on the shelves. Below was a wooden-doored storage section for pots, pans and large bowls. Years later, I treated myself to some pink cups and green plates like Pop's Depression Glass. The plates don't match the cups, but I often use the few teacups that remain or the plates when serving small portions of a rich, delectable, homemade dessert. The jug I have had for almost twenty years, a castoff from my sister-in-law who never had the attachment to it that I do and did not find it attractive. It and my collection of handmade pottery sit on shelves protected by the glass doors of my kitchen cabinets. We use the ceramic pitcher in summer for ice water or Mugicha, the cold, thirst-quenching Japanese roasted-barley tea that we love.

On the counter surface between the top and bottom, Pop kept a pad of paper, his fountain pen and some pencils next to the phone. I can close my eyes now, as I recall those days of my childhood, and see his spidery old-fashioned handwriting on the ruled writing pad. As a young school-aged girl, I wrote Poppy regular letters. His return letters to me were full of run-on sentences recounting the news of the community of family and friends, a report on his health, and from spring through fall, his excitement or disappointment concerning baseball. In high school, my salutation to him changed from "Dear Poppy" to "Dear Pop," and I told him how I was doing in school. When we visited, he would enlist my help in writing letters to Gilbert Seldes, a cultural critic and novelist, and his childhood friend. This successful Alliance boy wrote back to John Levin, his buddy from their early rural roots, each letter of which pleased Pop and made him feel proud.

The telephones of the 1940s and 50s are today's museum pieces. Large and heavy-bodied, a black Bakelite case held the inner works of bell, ringer and wires and was notable for its large dial with the numbers and letters divided and framed by finger-tip size holes. The caller placed fingers in the holes and rotated the dial clockwise while listening with headset to ear. In those days in rural areas, people shared lines with their nearest neighbors in a party line system. The issues of privacy engendered by the digital revolution present a completely different definition of "privacy." Other people on your party line could and did listen to your conversations. If you just wanted to call a friend or relative for a catchup conversation, it was rude to keep picking up the phone when you knew the party was engaged in a call. On the other hand, if you needed to make a timely or urgent call, repeated pickups could signal your need. If your hint was not acknowledged with a prompt end to their call, your recourse was to tell the person—someone you probably knew from the neighborhood—that you needed to use the phone and could he or she, please relinquish the line. Operators were necessary for calls beyond the party line and all long-distance services.

Telephone technicians came out to repair lines or actual phones. According to Wikipedia and earth911.com, which I have consulted on my Android-system smart phone, approximately 150 million mobile phones are discarded each year in the USA alone. For every million devices recycled, 35 thousand pounds of copper and 772 pounds silver are recoverable. The horrors of the contemporary waste stream and e-waste stream aside, only the aged population of today can appreciate the significance of repairing things rather than discarding them. In a process of elimination, one could walk or drive to a neighbor's to determine whether the problem was in the line or the individual phone. One summer at Pop's our phone suddenly failed to work. A short walk across the street to consult with old Mr. Berkowitz let us know it was our phone. When the repair man came a day or two later and opened the telephone, we found a large, female black widow spider, displaying on the underside of her abdomen the ominous, red hourglass. Pop had already taught us what they looked like by catching them and keeping them in jars until releasing them far from where we could encounter them—or squashing them underfoot. The stunning, graphic image of the natural spider stuck among the man-made parts of the phone interior has stayed with me all my life.

Although breezes came from windows in other parts of the house, the south wall held the only windows in the kitchen. They opened out onto the part of the yard where the largest, oldest trees stood and where family and friends gathered. A large enameled sink, counter tops made from planed, sanded and painted boards, and the small gas range in the corner completed the kitchen. If we revolve another forty-five degrees to the right, there it is again, the queen of the kitchen cabinets. This piece of furniture was a childhood friend of mine, inanimate, for sure, but a friend, so I turn again to it before leaving this beloved room of childhood. Also, before leaving the kitchen, we pause again for another cooling drink of artesian well water, and look out the window at memories of loved ones and experiences long gone.

The living room held a day bed or two, some chairs, including a horribly lumpy and uncomfortable one covered in worn maroon, cut velvet with spiral-cut wooden legs and upright pieces. There was also an old console-cabinet radio with its wooden and Bakelite facade, an old iron-work lamp—the identical twin of which a friend of mine has in her guest room—and a small, copper-lined, waist-high, wooden smoking cabinet on skinny legs. Opening and closing the door was such a pleasing sensation—the sound of the ball bearing in the catch making a gentle "pop," the feel of the delicate, pendant-style handle in my small hand, the shine of the copper lining—that I opened and closed it repeatedly, just for the sensation. I have seen similar smoking cabinets since then, just as I have seen large, heavy-looking vases with intricate floral relief that were carved from soapstone but looked like marble; such a one being a non-functional ornament in the room. I remember no art nor family photographs on the walls. An uncomfortable day bed—whether one sat on it or slept on it—occupied the north wall of the living room throughout most of my childhood.

The living room opened onto the screened porch where rocking chairs made from heavy wooden boards and twill-woven, splint seats lined up or alternated with sleeping cots. As we grandchildren grew, we all slept on the porch, and the whole family sat there to enjoy thunderstorms and gentle rain showers, inhaling deeply the sweet, refreshing fragrance known as petrichor, when after a drought, rain fell upon baked earth. My mother, who was a talented seamstress and knitter, had summertime crafts more suited to carrying with her and often kept her hands active on the porch with one or another of the projects with which I could assist her. For several summers, she and her friends and my aunts made purses from round, natural-color, split-rattan breadbaskets. They poked thin crochet hooks into the rims and crocheted round and round with a fine cotton or linen thread until they made a drawstring top for the basket, complete with strategically placed, brightly colored wooden beads. These bags were a summer fashion statement in the late 40s and early 50s, and I would love to make one now, but crochet is one needlework skill I never mastered, and my arthritic fingers are better employed with sewing and gardening.

Whenever we ate cantaloupe or honeydew melons, Mom or my aunts would thoroughly wash the seeds and dry them on a screen. Once dry, they spread out the seeds, and gave me a sharp needle and length of thread to pierce the seeds and string them on the thread for seed necklaces. Sometimes extra wooden beads from my mother's projects would find their way onto the leis of seed necklaces. When young, my brother and boy cousins also enjoyed this repetitive endeavor that yielded attractive, free adornments.

A notorious episode from the early 1950s has stuck in my memory all these years. My two boy cousins and I were there with my Aunt Esther and Uncle Harry in residence as the parental authorities. Pop was somewhere on Gershal Avenue visiting local friends, and my aunt and uncle were possibly in Vineland with cousins and friends for dinner out or a party. We were instructed to turn off the small black and white TV, mostly dominated by static and snow that made viewing whatever shows were on then less than satisfying, at a specific time and go to sleep. But Pop and aunt and uncle stayed out late, so we stayed up late. We got hungry and opened a can of Spaghetti-Os which we heated and took out to the porch to eat on flimsy paper plates. At some point we thought we heard a vehicle pull into the driveway and all ran for our cots. It was the hapless Bruce whose plate of canned pasta flipped and spilled all over his sheets. Either it was a false alarm or Pop, who came in and, seeing the porch dark, went straight to bed. We worked at lightning speed to strip the soiled sheets and stuff them into the laundry hamper and were sound asleep when the adults in charge returned. That morning, when my aunt began the laundry, she discovered our pathetic ruse and busted us. While the reprimand was seriously delivered and rationally explained, the punishment was not severe. I can still see a slow-motion version of Bruce's poorly executed leap into the cot and the plate of Spaghetti-Os flipping and splattering.

There were other times when my parents were out with relatives and friends and I was old enough to sleep on the porch with Pop at home, or my brother babysitting, when I could not sleep. I had a tendency towards anxiety and insomnia, the fear that something would happen to my parents and I would be orphaned. I'd listen to the cars driving past Pop's thinking each time I heard an approaching car that the next one would be my Mommy and Daddy. I was certain that it was much later than they usually stayed out, that something had happened and succumbed to worrying until I finally fell asleep or heard the car wheels pulling into the driveway and feigned sleep as they peeped into the porch to check on me.

We used the front door when we played in the front yard which, like all the land other than the leased fields, the small garden patches, the house and outbuildings, and woods rimming the expanse of the property, consisted of a grassy lawn planted with every kind of tree that would grow in South Jersey's sandy soil and four-season climate. A few steps led from the porch to the lawn, and lawn separated us from the road and the rural mailbox with its red metal flag, that we kids all wanted the privilege of checking and, if we were lucky when it was our turn, from which each loved being the bearer of its contents. We played catch and softball and other games on the swath of grass adjacent to the drive, but there was a spot where brambles grew—possibly blackberries; I only remember annoying thorns in my bare feet but no fruit. There, too, in a sunny spot, my mom would fill a galvanized tub with water in the morning to warm all day, so that on the days we could not get to the river or one of the lakes for special occasions, I could splash around and cool off. The next day, the water could be used for a young tree, vegetables or the sunflowers Pop planted.

The sun getting low in the west meant it was time to go into the house again. If I had played and gotten dirty and not "gone swimming" outside in the galvanized washtub, a bath would have been in order. The bathroom was such a simple affair that in a moment when words seem over the top, I am inclined to draw a little graphic representation of the room with a few geometric shapes to signify the functional components of that room. Remember, this is the 1940s and 1950s, the simple country home of a man set in his ways who lived alone and kept his possessions orderly, cared for, and to a minimum. The claw-foot tub occupied almost the full length of the bathroom on the right wall with a shelf that held eye wash and the deep, cobalt-blue, glass eye cup, over-the-counter medicines like Calamine lotion for poison ivy and mosquito bites, witch hazel, shampoo and soap, brushes, and sundry male toiletries, including his razor, shaving brush and wooden bowl, with a round cake of shaving soap. Towels hung from cheap chrome-plated, thin towel racks and hooks, and other shelves held the overflow of personal hygiene products to serve the needs of the family members in residence. The tub displayed a stain from iron in the water flowing through copper plumbing parts.

The one item that in comparison to country living seemed a luxury in our city dwelling was a roomy bathtub with a great shower. The bathroom in the country had character but was basic. I always sensed the smell of the water, a natural, clean, chlorine-free smell, intensified by the iron and copper pipes. Rinsing a soaped-up, impatient young body or lathered head was accomplished with the coral-red, rubber hose that attached to the finger-thin faucet of the tub and a rubber and tin shower head that attached to the other end. Frequently, the head fell off the hose or the hose fell off the faucet, and if that was not enough to make bath time challenging, often the well would need to refill. When that happened, the water from the rubber hose slowed to a trickle. The inconvenience of this phenomenon when I was a child was bad enough as I sat in the tub shivering while the water cooled, but in my teen years it reached crisis proportions when, after being at the beach all day, I was trying to bathe and get ready for an evening out with friends and cousins.

The toilet faced the door and spared little space between itself and the tub, which sat under a deep shelf at its front where additional towels and washcloths were compactly and neatly folded. At some point in my younger years, the toilet bowl seal had begun to leak and the wooden subfloor under the linoleum began to rot. For a while, toadstools grew from the floor, quite an amazing blending of a forest and a bathroom floor. Pop must have hired some young carpenters to make the necessary repairs during one school year because to my surprise—a mix of relief and disappointment—the fungi were completely gone within a couple years. As unsound as I realized this condition was, it was still fascinating to observe such a natural phenomenon of rot proceeding on the bathroom floor.

By the time my mom got me into my light, cotton pajamas and ready to sleep in the cot that occupied a corner of the second bedroom—the bedroom of the daughter of the week or month and her family—the sun was a red ball slipping behind the treetops at the boundary of the five acres. Also in that room was the old treadle sewing machine, another artifact from a simpler time and my family's past that I loved to operate by treadling with my feet, turning the fly wheel with my small hand, and searching through the small, long drawers.

When very young, I loved to have my mother read a few pages to me and later enjoyed reading myself. At some point, we had a record player and a handful of record albums and singles. One that meant a lot to the three generations when I turned nine was an album or a single by the folk group The Weavers: Pete Seeger, Lee Hays, Ronnie Gilbert and Fred Hellerman. Controversial for their political views at that time, they did covers of great roots, blues, folk and other music. Pop bought the record because of the hybrid Yiddish-Hebrew folk song "Tzena, Tzena, Tzena," but I also learned the Huddie "Lead Belly" Ledbetter song, "Irene, Goodnight" and several other songs that had me well enough versed to immerse myself in the folk revival of the latter 1950s and 60s. My head hit the pillow, covered by the cherry or flower-patterned cases my mother had sewn from the muslin flour sacks in which she bought large quantities (probably twenty-five pounds) of flour. My eyes unfocused as I stared at the striped and flowered wallpaper and listened to the plaintive sound of the mourning doves. Soon I was asleep, dreaming of exploring things in Poppy's cellar the next day, getting Mommy to let me sew on the treadle sewing machine under the bedroom's north window, or going to the river.

I realize as I recall my treasured artifacts of memory that huge gaps exist in my mental storehouse. I recall that linoleum covered at least the bathroom and kitchen floors, but no fading, patterned outlines remain. No bright colors jump out at me. No doubt the linoleum was scuffed by the feet of the four grandkids and then the fifth born in 1952, the adults of the family, and countless relatives and friends, not to mention the dogs' claws as they ran into the house for protection from thunderstorms. Yet recently, in Salida, Colorado, while visiting a friend, the colorful spatters and spots of a patch of worn linoleum in an old part of her house which connects to the newer additions rang a bell. Sometimes in life we encounter something that brings back an amorphous memory, something we cannot pinpoint with laser sharpness but that evokes such an old familiarity that we feel we have returned to a deep part of our long gone home.

∿

John Levin's Domain, Bedroom and Basement

Pop's bedroom was spartan. His double bed was always neatly made, as if he had been trained to make it in the military and to tidy his quarters and continued the habit the rest of his life. In fact, he never was in the military. Although his age—early thirties—would have made him a prime infantryman, flat feet, poor hearing and vision, and, most important, his wife and three small daughters kept him out of World War I. The bedding on the metal-framed double bed, consisting of a Bates brand, jacquard-weave cotton bedspread and a Hudson Bay woolen blanket, was always taut, neat and clean. There were few or no closets in simple bungalows in those days, so armoires or what were called chifforobes (an armoire or wardrobe—a place to hang things—combined with drawers on the opposite side) housed clothing. A large, mirrored one took up the short wall opposite the foot of the bed, and I believe a compact cedar chest on the left wall constituted the furniture collection in the crowded bedroom. Sometimes my brother or a cousin would sleep in Pop's bed with him when the house was full of family; unfortunate lads they were, because Poppy's snores thundered and snorted like the fiercest horses of mythology. Pop built his own night tables, one on each side of the bed where the windup alarm clock, his eyeglasses, a glass of water and other essentials were neatly arranged. There, too, may have been another glass of water with his dentures, but although I remember the strangeness of teeth in a jar, it seems more likely that they would have been kept in the bathroom. His green, twill-weave drawstring pouch, in which he kept his egg route money, would be placed in a drawer. The east window next to his bed allowed early morning light to enter the room and waken him naturally, and on the wall shared with the bathroom stood a dresser and mirror. Wooden boxes held some cuff links and tie pins which I don't remember him ever wearing—although certainly he appeared in a suit and tie for the bar mitzvahs of my brother and cousins and my brother's wedding—and a few arrowheads or other found treasures. Current issues of *FIELD and STREAM* were stacked a few high on the night table or dresser, later accompanied by an occasional magazine that extolled the virtues of a natural lifestyle à la nudism, with titles such as *SUN LORE*. The walls were probably papered in something with subtly hued stripes, but that eludes me. A small, woven throw rug lay next to his bed.

But the real male domain, the space in the bungalow least "invaded" by my mother and aunts, whose time spent for weeks in summers was both desired and depended upon, was the basement or the cellar. I loved going into his domain as much as my brother and cousins but was intimidated when young and small by the five, open-backed steps with no railing. Once bravely at the bottom, I was among plentiful and pleasing rewards.

For years when I was very young there was a storage barrel made from heavy plied-paper with a clipped, metal rim to close the lid. Somehow, I got in there, whether my persistent, nimble hands were strong enough to open it and find treasure inside to titillate my curiosity or because someone else had gotten in and left the lid ajar. Something glittered, a rippling mass of foamy, light-green crepe, beaded all over with crystal-colored, glass bugle and seed beads and embedded with rhinestones. Its heft surprised me as I strained to pick it up because it sparkled like sunlight on the seafoam of crashing waves at the beaches of Absecon Island—Atlantic City, Ventnor and Margate—where my father's sisters lived. It was a gown, surely something my grandmother had worn in the distant past, and I developed a desire to pry open the tiny prongs of the little, metal rhinestone settings and collect the faceted glass. I remember doing this for an unspecified period that, to my young mind, seemed to extend for several years. It was probably a few times one season, and my little fingernails could not have removed many. Did my child's mind think they were diamonds? I don't remember being punished for doing it, although I have long realized it was a destructive activity, albeit prompted by curiosity as is the desire to take apart a gear-driven watch or machine. The gown seemed to engender no great emotional attachment in my grandfather or my mother, and it was possibly already dry rotting, the way long, but improperly stored fabric eventually will. At some point during another trip to visit Pop, I became aware that the barrel was gone, but I can still feel the weight of the dress, not from the beautiful drapery of light-celadon crepe but from all the bead work and shining fake stones clamped onto it. The color is one I still love, and although I don't generally wear clothing with faux jewels and sequins, I sometimes indulge my delight in sparkle to this day with earrings or necklaces that glitter like ice or sun on seafoam.

That dress and the other contents of the barrel, however, were the only truly feminine items in that cellar, a large expanse that filled the space under the entire house except the screened porch. Although it was a man's realm, it was, nevertheless, neat as a pin and clean enough to eat from all its surfaces. Nothing about my house and homestead is anywhere near as neat and as clean as my grandfather's house, his cellar and his outbuildings were.

In the corner on an elevated platform stood the wringer washing machine, close to a window, with a hose attached to save the grey water that emptied from the machine in buckets and tubs to irrigate young trees. Near it was a washtub for serious clean up jobs. Then came the shelves and tools, hand planes in several sizes, chisels, hammers for different purposes, saws of every type, straightedges, T-squares, folding measuring tools, and levels in several sizes, their floating bubbles an intriguing puzzle to me. Empty baby food and pickled herring jars held scrupulously sorted nails and screws, and tins of every size held more delights for the curious child in awe of the maker and of the material parts it took to make things of wood and metal, things with significance and precision. There was a section with chains and ropes and anything imaginable, big stuff in the corner. Other shelves in safe spots stored surplus canning and gallon jars for Pop's fermented cucumber and green-tomato pickles and kitchen equipment not regularly used. When Mickey was a young teen and I was six or seven, Pop gave us free rein to craft me a small, wooden boat for my bath. My impatience and Mickey's interest in other things curtailed the time we spent at this pursuit, and the boat was nothing of great skill, but I long loved its simplicity and provenance in my bathtub.

But the very best thing along the basement's south wall was a small device that we craved permission to use: the egg candler. In my first decade of life, it was stored well beyond my reach on the shelf in front of a light bulb, which was switched on and off with a pull string. A ringed cup affair held the egg so that the light—originally a literal candle—would shine and reveal, x-ray like, any blood spots in the egg. That indicated that the egg was fertilized and not acceptable for those who observed kashrut or kosher laws and therefore not to be sold by Pop to his customers in Philadelphia as kosher eggs. I don't know what happened to the eggs with tiny blood specks, but we probably did not eat them. That decision was more a cultural bias rather than a religious observance because none of the sisters nor their father kept kosher at that point. If the blood spots were very small, he probably sold them to non-Jewish neighbors. He got the eggs wholesale from any one of many kosher poultry farms in the area of South Norma.

Other members of the greater Levin family who lived in the country—Uncle Bill and Aunt Lena and their country sons—had painted-tin egg scales, another of those basic farm-household utensils that for me still exudes simple charm. I would feel a little sentimental and silly admitting this except for the fact that I am, after all, the kind of poet who writes an ode to a potato masher. I am in excellent company with Pablo Neruda, the late Chilean poet, who wrote a book of poems *Ode to Common Things*. When I saw the same tin egg scale in the home of my mother's first cousin and his wife, now owned by his grandson and granddaughter-in-law in August of 2018, I instantly knew it as a part of that family's artifacts of their agrarian past. Perhaps there were several of them, even one at Poppy's.

One of my earliest memories in Philadelphia, from about age three and a half or four, is closely connected to Pop's egg route. On a partly cloudy afternoon when Pop came to

119

our house with eggs, Aunt Lena's sour cream, and possibly chickens from one of the farms, he lay down his green cotton, flat-topped cap and spilled his coin purse on the dining table for Mickey and me to take some coins. At that moment the sun emerged from behind a cloud and shone through the lightly frosted window of our corner house to catch the event in an eternal, automatic-flash, candid photo. Somehow that memory is deeply conflated with candling eggs for his little business enterprise.

The northeast corner of the cellar was walled off as a coal bin which Pop had filled prior to cold weather via a chute from the delivery truck to the basement. He must have required frequent coal deliveries to keep even a couple rooms warm in winter. The furnace seemed large in its central position of the concrete and cinderblock cellar. The furnace could keep a fire all day or all night once the fire was built and properly banked. For me, getting out of bed in the morning when we stayed overnight in cold weather was torture, but any extra visit above and beyond our regular schedule was always a special treat. Besides, Pop or my brother Mickey or my dad, who fed the coal furnace at home in Philadelphia through the 40s, was responsible for the furnace, not a little girl.

Nowadays, for a few nights a week in the cold seasons while my husband is at work all night in his job as caretaker of a developmentally disabled gentleman, I am responsible for keeping fed the two wood stoves that heat our house. I still don't like to get up on a cold morning when the coals of the night's fire are all that remain. I pull on robe and socks and tend the fires that soon come to life from raked embers and fresh wood. The activities of moving around, feeding pets and breathing fresh, outside air get my blood flowing and warm me. My stiff joints start working smoothly as endorphins kick in and I wake, alive and joyful to a new day. But the wood stoves of the long, extended second half of my life are a far cry from the sometimes-balky, dirty coal furnaces of my childhood and my grandfather's cellar.

In another section of the room, huge winter galoshes with buckles requiring adept fingers and strong legs, rain boots and waders, rainslickers and hats, and things packed in odd-shaped boxes of heavy, plied-paper with metal reinforced corners transitioned to the woodworking area. Stacks of lumber and woodwork projects (bookshelves, benches, simple cabinets) spread across an out of the way section of the floor in various stages of construction. He kept his cans of paint well cleaned and carefully arranged on shelves. There was a period when his well-made, but basic furniture was painted instead of varnished or shellacked, as was the book case of pine boards he made for my brother's bedroom wall.

He had learned the faux-marble finishing technique, as old as cave painting and employed in classical periods as well as in the neo-classical resurgence of the nineteenth century and the Art Deco era of the 1920s. Taken to its height, the faux technique became trompe l'oeil and gave incredible three-dimensional qualities to the two-dimensional art forms of painting and drawing. Pop was utilitarian and far from artistic, and his infatuation with the ancient and revived techniques never went beyond feathery strokes of white paint applied to a base coat of pale-green paint on wooden surfaces, with the intention of fooling the viewer into thinking that he or she was admiring a simple table carved from marble. His meticulous work habits and love of making reliable, pleasing goods earned him extra income from this period of his hobby. His father, after all, in his young adulthood had been a cabinet maker in the service of the Russian government. Since Israel Hersh Levin left nothing behind by which we can judge his skill, his artistry is a moot point. Pop would never paint an interior scape that created a sensation of being in a long room with vaulted ceilings and ornate draperies; nor did he ever craft fine furniture. His art was in living an authentic life, in being his own fully realized character. In my infancy, Pop had commissioned a highly skilled woodworking relative, whose name I cannot recall, to build a beautifully crafted doll cradle. Pop painted it, and my mother made the bedding for the dolls. After my childhood, I used it for magazines. It perished in the house fire that destroyed so many photos and artifacts that my husband and I treasured from our individual pasts and first days together. Although he did not build it, it was something I always associated with my grandfather.

On one wall near the coal bin hung shovels of all kinds: coal, snow, garden; adzes, axes, hatchets and mauls cushioned safely in crates. Near the shovels hung the Flexible Flyer snow sled, a perfect symbol of childhood freedom. Nothing signified the glories of a childhood winter memory better than flying along on a snow sled, steering my own destiny on a smooth, exciting glide, whether my grandfather or parent or older brother was pulling me along a flat road or an older version of myself went zooming downhill on my own accord on the wooden bed of a steel-bladed Flexible Flyer sled.

∾

The Animal Kingdom at Pop's Place

Tiny puppies, whose eyes had already opened and who had begun to crawl, constitute some of my earliest and most innocent memories of dogs. Pop entertained us by swaddling the pups in his white cotton socks and securing the socks with wooden clothes pins to the rope of clothesline. The puppies were snug and safe but after we photographed them, we took them down and set them free again. I doubt that this would qualify as animal abuse even by today's standards and know that had the puppies squealed in pain or fear, our cries in defense of them would have ended the practice. I still have a couple photographs of puppies in a row of socks on the line and one of Mickey and me cradling their squirming, soft, sausage-like bodies. In my very youngest years, Poppy had the best dog in the world—or so we grandkids thought. Mitzie, a purebred white Spitz, was well behaved, affectionate, protective, fluffy, and smelled lovely after her baths. She was, however, terrified of storms and ran between the legs of whichever of her humans was ahead of her in the mad dash for the cover of the house when a sudden storm blew in. Like any dog, especially ones with light colored coats, she felt the onus, after a bath, of restoring her coat to what must have felt more natural to her. After shaking herself free of excess water and getting us wet in the process, she went promptly upon release to her favorite spot and rolled in dirt that was well impregnated with motor oil from a leaking car engine. Pop could not control his frustration and cursed a blue streak, his favorite imprecation being "Son of a Bitch!"—in this case a species-correct but a gender-inaccurate epithet. Once inside the house, she curled up on linoleum or old pine-boarded floors and slept.

Mitzie, the favored dog, the progenitor of all our canine dreams, but only one of many to follow, met a tragic end. I must have been about nine when it happened. Pop was in the hospital in Philadelphia, and Mitzie was staying at our house. Mickey, at fifteen, was contemplating veterinary medicine as his field of study after high school. He adored Mitzie, of course, and took her on morning, afternoon and evening walks every day. One evening, she slipped her leash or managed to run into the street. She was hit by a car, the driver of which felt remorseful and took them to a nearby vet. The vet could not save her, and although we all grieved deeply, Micky felt an undeserved guilt. I think that that experience opened his eyes to the emotional difficulties he would encounter when failing to save suffering animals. That, and the fact that he met his future wife, Beth Abrams, when they were both young, influenced his decision to study pharmacy instead.

A long line of various terriers followed, but they were not beautiful and sweet; the hyperactivity of terriers still convinces me that cats, not dogs, are the proper pets to snuggle on a human lap. One named Misty was so unlike the Spitz whose name could easily have been mistaken for hers. A boxer named Peggy that Pop adopted was gentle and smart, but her size intimidated me. At one point an Airedale became the breed of choice, and Duke's temperament, liveliness, and curly coat charmed me. Since every dog I have ever had as an adult has been a mixed-breed rescue dog or a "please take one-of-the-litter" rescues, I have never had an Airedale or Spitz of my own.

My father worked in the wholesale food business, first in the Dock Street area near the Delaware River and then a huge Food Distribution Center in South Philadelphia. Now and then he came home with a stray dog in tow, which we kept at home until our next trip to visit Pop. Many a lost or abandoned city dog passed through our conduit to Pop's. He rarely kept the dogs but found homes among family members or neighbors. Most country dwellers have a need for a dog or two, but my parents were firm believers that it was cruel to keep a dog confined to a house in the city all day. So even the most delightful pooches that found their way into our hearts and home were only in transit. One handsome hunting dog—a Black and Tan I later learned—was, for a while, our foster dog. We made a bed for it in the basement and walked and cared for it conscientiously; however, that dog turned out to be a snake in the grass. I must have been about eleven or twelve and a little old to still be playing with dolls, but my collection of Madame Alexander dolls—Meg, Jo, Amy and Beth from *Little Women* and another doll dressed in beautiful Victorian clothing—were arrayed on a bench that Pop had made for me in the basement. That dog chewed the feet of each doll, ruining their value and my joy in them.

Had the dog been allowed in the living room, it would have chewed the legs of all the furniture. It chewed my pointe shoes just a couple days before the Sunday ballet class I took with the great British choreographer, Antony Tudor. Tudor had emigrated from England to New York where he was closely associated with the American Ballet Theater in Manhattan for the rest of his life. On Sundays from about 1952 to 1954, Tudor took the train from New York to Thirtieth Street Station in Philadelphia and then the Frankford Elevated Train to the neighborhood dance studio where I took ballet lessons during the week. It was a rare opportunity for girls in my age group to study with him, and my parents and I were honored to give up our Sundays for the chance to have me study with this great man of twentieth-century ballet, even when it meant altering the routine of our visits to Pop's. First, he taught a class of adults—professional dancers and dance teachers—and then our class of adolescent and young teen girls. He had an impressive British accent and the austere demeanor of a grand ballet master, which was more a dramatic persona than his real nature, I think, because he acknowledged the unique talents and personalities of his young students and had deep affection for us despite the theatrical embarrassments he sometimes concocted for us if we yawned or slouched in class or seemed to be going through the motions rather than being passionate about dancing. I loved the first hour of the class when we did barre and then floor work in our soft, kid-leather ballet shoes, but as beautiful as I thought dancing on pointe was to watch, it was extremely painful for me. So, one Sunday when I showed my savagely dog-chewed, pointe shoes to Mr. T. and explained that the dog had eaten them, everyone laughed. I was sheepish and embarrassed but grateful for the quirky version of the cliched excuse that "the dog ate my homework." That dog went to Pop's as soon as we were able to get there, and I was not sorry to see it go.

Whenever Pop was hospitalized—a more frequent occurrence by the time I was in high school—we took care of whatever dog he had at the time. There was a nervous miniature Doberman Pinscher with two little pups that required regular walking. By now Mickey and Beth were married, and my mom, dad and I shared canine care. Mom clipped the dog's leash to the lid of the trashcan in the driveway behind our house as she took out the trash and garbage. Something spooked the poor dog, and it knocked over the can and took off down the driveway, trailing the trashcan lid. Mom's hot pursuit was unsuccessful, and despite thorough inquiries the dog was never seen again. Although it required an extra half hour in the morning, I was delighted to lose the sleep and warm a milk replacer, fill the bottles and feed the hungry little puppies. My girlfriends were just as thrilled to help with the afternoon feeding after school. Worn down by my entreaties to be allowed to keep one or both, my weary parents gave their consent. I think, however, that upon release from the hospital and return to his home in the country, Pop wanted the pups back as consolation for losing their mother. Clearly, the urban Weinsteins were not fated to enjoy a family dog.

The menagerie was not limited to dogs. Sadly, though, there were no cats ever at my grandfather's. My mother feared cats, having been badly scratched and bitten by an ill-tempered feline that lived at the home of my father's youngest sister. Mom's dread of cats, however, must have stemmed from something closer to home, and I seem to remember that Pop hated cats. It took some years for me to shake the family distaste for cats, but I am comfortable claiming the mantle of black sheep, and despite their predatory behavior, I love felines. Even our cats are rescues, and along with every dog, male or female, the cats that find their way to our home are spayed and neutered as early as possible.

Early on there was an ever-changing parade of fancy poultry from squab—little pigeons, which they cooked to tempt me to eat a protein and fat rich meal—to guinea hens. I gratified my grandfather and parents, who all worried because I was so thin, by gobbling up vegetables and the juicy roasted squabs, which they called "little chickens," but then refused to eat it again once I learned that it was one of the little birds I had held. Fancy chickens followed. I also have a vague childhood memory of peacocks and peahens passing through. The memory is like the flash of the male bird's phenomenally fanned tail. There and then gone, as loud and attention getting as the shriek of the bird's cries, but less than warm and cuddly, an attribute for which John Levin's three daughters and their children had a predilection. I know that an experimental cavalcade of familiar and nearly exotic creatures came through that little homestead over a period of close to twenty years until it all began to wind down.

At some point there were rabbits in hutches, but I can only remember trying to hold them and finding it too difficult to be enjoyable, feeding them carrot sticks and lettuce, and stroking their amazingly soft coats. I have no memory of them on the plate, however. Knowing how rabbits scream upon being slaughtered, from the experience of friends who have raised them, it does not surprise me that my mother and aunts and all of us kids would unite and refuse to eat these lovable, vulnerable creatures once we had stroked downy fur. Perhaps, though, Pop occasionally managed to fool not just his grandchildren, but his daughters, who had become citified and ignored the menagerie as much as possible, into thinking it was poultry they were eating.

The stars of Pop's barnyard were the dairy goats. Few domestic animals provide as much delight as young goat kids frisking and leaping everywhere. From the ground to a precarious, high perch in one balletic move, they are perpetual motion machines. They smell milky from the sustenance they suckle from their mothers' teats or from hand-held bottles and fresh from their straw bedding and the hay that is fed to the does. Goats have the highest mammalian body temperature, and snuggling a kid goat while feeding it is a joy for a child. Pop must have first bought goats when I was around nine or ten years old. I don't have a memory of a buck goat. Probably Pop took the doe—or nannies as we called them—to another farm that had a buck goat. Had he had a buck, I think I would remember the smell and overtly aggressive reproductive behavior of the billie goat. Buck goats in reproductive mode display mating behavior by lifting their upper lips to squirt and taste their own urine, a habit that would certainly have repulsed the finicky girl that I was, for years to come.

The spring in which I would turn eleven, instead of celebrating the Jewish holiday of Passover at our home, the three sisters decided that we would join a huge group of Jewish families still left in the Alliance/Norma community, and their relatives coming from the mid-Atlantic region, at the Hirsch Hotel Community Seder. Pop could not leave his lactating goats and their bottle-fed kids, and participating in the Hotel Seder, while expensive, gave the hard-working sisters a break from the cleaning and cooking necessary for the holiday. The huge dining hall was filled and loud. I remember the blue and white dress with red trim I wore for the holiday and the discomfort I felt through the whole Seder. I was beginning to develop breasts, and far from being excited about this proposition, I was miserable. Anything touching my skinny, still flat chest was excruciatingly painful.

Our parents planned a surprise with Pop. Mickey and I would be allowed to stay for the extended weekend at Pop's without Mom and Dad. That was a milestone, and I felt proud of my brother who would have the responsibility of caring for me, which he always did with kindness. In the morning, however, when I wanted milk from the store for my cereal, Mickey insisted that the goats' milk was all there was, and that Poppy would not drive to the store to get me the kind of milk I was used to and wanted. Mickey took charge and I complied. Being the food-finicky kid that I was, I did not like it at first. I got used to it because I loved the goats—even when they butted me—and wanted to please my grandfather.

When my husband Joe and I kept goats for thirteen years—first in Pennsylvania and then in Arkansas—nothing tasted better than the yogurt and cheeses I made from their milk. I staunchly defended the various cheeses against the rejections by friends who had what I deemed less sophisticated palates. When a friend with decided taste in food turned up her nose at my proffered goat cheese, I told her that her cat would probably like it whether she did or not. Like owner like cat. Her otherwise charming kitty raised her pink nose into the air and walked away from the pinch of home-made goat cheese with total disdain. Goat cheese is so much more popular in 2019 than in it was in 1982.

It was during the late 1940s, early 50s, when a procession of animals came and went, that Pop's five acres was most like a farm. The most exciting and yet disappointing was a Shetland pony. What child of that era did not dream of having a pony? John Steinbeck's *The Red Pony* (published as a complete book in 1937), the movie version of 1949, and *Misty of Chincoteague* (the 1947 children's novel by Marguerite Henry, about the wild horses of Chincoteague Island) were significant shapers of those equine-focused desires of childhood. The pony was iconic in appearance: short, stolid, shaggy but also possessed of an almost malevolent nature. It did not tolerate us being placed upon its back for rides and, even more than the goats, had the habit of taking everything possible in its mouth—from my aunt's wristwatch and hair net to our arms, if we tried to scratch its muzzle when it was not in the mood. I don't know how it had been treated before it arrived at Pop's, and he was not the kind of pet owner who is so prevalent today. He was neither overly indulgent nor abusive, but he did discipline animals with a rolled-up newspaper or mag-azine on occasion. No doubt dog and horse whisperers of today would be appalled. I cannot remember how long the pony lived at Pop's, but I do remember that its behavior that seemed so funny when it tried to eat my Aunt Jean's watch was anything but humorous when it left teeth marks on my skinny arm. On principle I mourned its departure while feeling relief that I did not have to pretend to love it. There was always another friend or acquaintance to assume responsibility for the failed experiments at Pop's.

~

Family and Friends Gather at John Levin's

My mother's photo album, long since dissected, scattered and largely lost, once held photos of my baldheaded, but fit grandfather in his perennial garb of khaki pants and blindingly white T-shirt, often with his arm around some young woman, either Laura Lehmann, who lived down the road, or another family friend. Sometimes Pop and the young friend posed theatrically in front of the Brownie box camera. I have a memory of a photo of a man pretending to be a ship-wrecked sailor in tattered pants standing behind a young woman, his arms wrapped around her. In my memory it is Pop, but then it morphs into my father. Either way, the likelihood of ever finding the photo to prove who the ghostly male was is slim. But miming for the camera was only one part of such joyous summer gatherings in the country.

There were often barbeques and picnics attended by our entire extended family and many friends. Pop had built a series of concrete block barbeque pits, simple as a pile of blocks and grates. The fire was safely contained, the grates could be raised and lowered to grill the hamburgers and Hebrew National Kosher hotdogs, and there was room for a large pot in which to cook corn or space to grill the corn before the burgers and hotdogs were finished. My aunt Esther's macaroni salad was a favorite. My mother or father made the potato salad, coleslaw and sliced tomato salads. Although my father was famous for what he called Greek salad, that was usually his colder weather contribution to non-traditional holiday meals, but sometimes he prepared it for summer picnics. There was nothing remotely Greek by today's paradigm of a classic Greek salad; rather it was, as a chef of Jewish background and roughly my age once explained to me, a "Russian" composite salad attributed to Greek origins. It contained everything but the proverbial kitchen sink: cabbage, potatoes, herring, anchovies, olives, beets, peppers, chunks of kosher salami, onion, kosher dill pickles, and other vegetables cut into bite-sized pieces—all marinated in olive oil and wine vinegar. I'm certain that if I tried to reproduce it today, it would fall far short of my father's savory concoction. My Aunt Jean's oil-based Jewish apple cake (or a summery, stone-fruit version) would have been on the picnic table, as well as big slices of watermelon and cantaloupe for dessert.

The gaiety of the gathering was not fueled by alcohol, although perhaps the men drank a little *schnapps*—not the peppermint or flavored varieties of spirits sold today, but a shot of whiskey enjoyed in moderation. *Schnapps* was the generic Yiddish word applied to any basic alcoholic drink that was not a cocktail, wine or beer. Cocktails were something largely limited to weddings, bar mitzvahs and such dress up celebrations, not to picnics. Nor do I recall the consumption of beer as a regular part of Jewish social feasting, but I may have just steered clear of this strong-smelling, intoxicating beverage. We were all glad to be in the sunshine and fresh air of the country where the water from the artesian wells was sweet and pure. There were always more adults than kids, but we were somehow the center of the focus and indulged appropriately. We played softball on the lawn away from the food and picnickers or took turns in the hammock, only to have our cousins try to oust us from its stout canvas cocoon. I wrapped myself up in the vast expanse of the fabric and held tight, a chrysalis determined not to emerge until the right time. We climbed trees and played with the dogs and whatever we could find that had the capacity to keep us entertained for hours.

For four or five years, once a summer, the husbands of the Levin sisters would arrange a boat charter from the seashore about forty miles east of Salem County, and in a couple of cars, the men and boys and one or two adventurous women would leave very early in the morning with the heavy, insulated, tin iceboxes, an early version of today's coolers, to carry back their haul of fish or seafood. I remember the crab catch more than fish, the fishing probably being more of a sport than something for the pot. However, none of the men in the family ever had a stuffed and mounted fish adorning the living room wall of their modest rowhouse or apartment. I never got to go on these outings because of my tendency to get carsick. I pleaded and promised that I would not get sick, that it wasn't fair that I never got to go, that it wasn't fair that I was born a girl—all to no avail. No one wanted to sit near me in the car on the way to the shore or back, and certainly no one wanted to deal with my dry heaves on a boat that rocked with the waves. Although I have long understood and empathized with their desires for a carefree holiday, it took me years as a youngster to forgive them for excluding me from the grand adventure.

By the time the old salts returned to the Gershal Avenue farmette, the dishes that accompanied the crab boil were prepared. The men dumped the crabs on the grass for the entertainment of one and all while the large pot of water heated on the grate of the concrete-block barbeque pit/grill. The blue-carapaced shellfish scuttled around the lawn while we kids kept them from scurrying away with sticks. I was fascinated by the mottled, turquoise-blue shells covering their bodies, their strange, antennae-like eye stalks, and the many legs and reddish-tipped claws with strong pincers. I would always try to pick one up then squeal in fear and drop it as it tried to pinch me in self-defense. My boy cousins, who had helped to catch them and knew how to handle them, delighted in frightening me by chasing me around the yard brandishing a carefully gripped crab that was frantically moving its claws in its effort to free and defend itself. Soon, though, each crab was collected and plunged with seasoning into the pot of boiling salted water. I did not exactly feel great sympathy for the things that threatened to sharply pinch my bare toes or fingers as I scampered among them and tried to handle them. Their claws drew blood. I did not then feel sorry that they were boiled to a painful death, and I was amazed by the metamorphosis in color they underwent in the pot, but I was still a fussy eater and was put off by the strong smell. My parents probably tempted me to just taste the choicest morsel, which I did and then turned up my nose and ate an ear of grilled corn slathered with butter or a juicy, sweet slice of watermelon.

ᔑ

Visiting Neighbors on South Gershal Avenue

My memory of the Hellmans, who were older than Pop, are amorphous. I think they were among the Holocaust survivor families but may, instead, have been one of the German families who arrived in the community in the 1930s in a prescient uprooting of their comfortable, worldly lives in Germany. Trying to pluck out details of their lives and close friendship with my Poppy now, I have the urge to turn to my brother or my parents, even Pop himself—all of whom seem so solid and alive in my memory—and ask them to tell me more about the Hellmans. Even stranger than the nearly palpable vestiges of my long-gone family is the fact that when I was in that community in August of 2018, I pointed out the house that was theirs so long ago as my new friend drove me around to revisit the old haunts. If they had children, they must have been gone from home by the time I interacted with the neighbors, but I can recall the sense of being small and overwhelmed in their huge house filled with their dark furniture and old-world possessions. I seem to remember that they always offered a little treat, chocolates or cookies, to the children and a small glass of schnapps or wine to the adults so that they could lift their glasses and drink to life in the traditional Hebrew toast: *L'Chaim*!

Into the early 1950s, the house of Sam and Sylvia Berkowitz and their three daughters—Carol, Toby and Debbie—stood one or two houses down from the Hellmans. The Berkowitzes, like my grandmother's family and so many others, were among the second wave of immigrants and were close friends of my parents. The two Sylvias looked alike, with dark hair, dark flashing eyes and smooth olive skin. They were beautiful woman who had traded the stylish dresses, faux pearl necklaces and cloche hats of their days as young flappers for the cotton print dresses and aprons of wives and mothers. Sam and his brother George owned the feed mill in Norma. One of George's daughters married the son of my mother's first cousin on her mother's side. The generations of the settler families were so close that, in my childhood, I believed we were all related, all braided together into some intricately woven and looped community of shared grandparents, aunts and uncles, cousins and friends, twisting around each other in a dancing, double helix of community. Genetically, all Ashkenazi (Eastern European Jews) are thirty-second cousins, and we all shared the same cultural markers as well as religious heritage.

The Berkowitz family featured large in my early childhood. The friendship between Sylvia Berkowitz and my mother could have been a model for a TV program. They were beautiful, vibrant woman who enjoyed life and friendship and loved their families. Their relationship blended *Ozzie and Harriet, I Love Lucy* and *Leave It to Beaver*—the TV shows of my childhood and young adulthood. I have memories of security and a loving home life based on things like two women friends canning peaches and tomatoes, fun outings to our usual swimming hole, the Norma Beach a mile away on the Maurice River, or to a large lake in Centerton where a sliding board (so tall I grew nervous climbing it) deposited us in the lake, in water that was deep enough to land in safely but not over our heads. The combination of fear and thrill always spurred a child on to more adventurous play, and supportive playmates lent courage to the laggards. This lake was where we sometimes went for outings like special birthdays. It was a more developed "resort" with appeal to a wider community than just the Jewish descendants of the Alliance Colony and their friends and families from Philadelphia, New York and other places farther away. We always packed huge big picnic lunches when we went there but left room for Dixie Cups of ice cream or popsicles.

There were occasions almost as zany as those shenanigans Lucy lured or coerced Ethel to participate in, when Sylvia Berkowitz would pick my mother up at Pop's in one of the huge, flat-bed trucks used at the mill for deliveries if she had no access to smaller vehicles. My mother did not learn to drive until I was older and never developed confidence as a driver, but her friend who was as petit as my mom hoisted herself into the driver's seat as my mother willingly climbed into the passenger seat, nearly doubled over with laughter, and off they went shpatzeering. *Shpatzeert* is a Yiddish word that literally means "going for a walk," but with the addition of "ing," the English present participle, it develops more layered connotations like "gallivanting" or "going out on the town," walking or going about not just for some fresh air, but to be seen or to participate in a fun, socially-oriented activity. Their favorite child-free way to *shpatzeert* was to go antiquing, which they were only able to do on an occasional Saturday when my father came in from the city and took over child care duty or when Pop would take my brother and me to Rainbow Lake or Parvin State Park or to visit his friends, the Stavitsky brothers.

These lakes are part of the watershed of the Muddy Run and have the typical, tannin-rich, reddish-brown water of South Jersey rivers, which we called "cedar water." We had boating and fishing outings with Pop at the lakes, renting a rowboat at Parvin's and casting fishing lines into the water, watching for concentric ripples of water in the small lake that meant the fish were biting. While we were exploring the countryside with Pop and our dad, my mom and her friend were scavenging the many antique and junkshops in a twenty-mile radius.

The two Sylvias came home with oil paintings or hand-painted china, Bohemian glass vases, or small pieces of furniture. I received, maybe fifteen years ago, a small, German porcelain vase of my mother's that my brother's wife rejected. My décor is eclectic, filled with many hand-crafted objects from pottery to textiles to paintings to glass, but other than the hand-craftsmanship, there is no unifying decorative or stylistic theme. Our house is rustic, and some would say that the porcelain vase is not appropriate for my house. This small vase has four, fragile gold-leafed feet and little handles on its rather rounded body. It is exquisitely painted with a poesy of rose- and violet-colored flowers and green and gold leaves on its front below the small, flared neck. I loved it when my mother bought it all those years ago on an antiquing jaunt with her friend, and I love it now when I fill it with spring flowers and place it somewhere off limits to climbing cats so that it may bring back those sweet memories of my young and beautiful mother.

The three Berkowitz girls were Carol, the same age as my brother Mickey, her middle sister Toby—who lives forever in a tiny black and white photograph with my brother, my aunt and cousin and me—and their younger sister, Debbie who was a couple years younger than I. While our mothers were in the Berkowitzes' kitchen canning or baking pies from local, seasonal fruit, we played in the girls' bedroom with dolls and cutouts, coloring books and tea sets. They had, in my earliest years, a record player, the kind that played one record at a time. Its date preceded the smaller record players that spun the 45s of a few years hence, but the machine had a Mickey Mouse finial that screwed on to the top of the spindle to secure the record in place. For years, certain musical rhythms, though I cannot possibly remember the tunes, reminded me of myself listening and watching—totally mesmerized—that figurine of Mickey Mouse revolving continuously as we played the records. It seemed to me that the little ceramic or Bakelite Mickey Mouse was pirouetting over, and over again, but out of sync with the meter of whatever music was playing. It was a strange awareness for such a young child to have, and it stuck with me for years. In my late twenties or early thirties, when despairing about the unsatisfying life I was living, I wrote a poem employing that image and using the sense of a grating, off-key meter and motion as a metaphor for my unhappiness. This sensation and the inchoate awareness of it was not, in my childhood, an unhappy feeling, just one of the silent observations some children make and never completely understand at the time. The times with those girl playmates were wonderful for a girl whose closest relatives of her generation were all boys. Trading clothes and receiving the hand-me-downs of the two older sisters was always a delight for me. As we got older, the differences in our ages mattered more, and Toby at eighteen to my sixteen had moved beyond me socially while Debbie had not yet caught up.

   After Sylvia and Sam moved into a newer, more modern house in Vineland than their large and rambling farmhouse on Gershal Avenue, Sam and Bertie Lamin bought the house and became close friends of Pop's and the three sisters and their families who spent summers and visited on weekends. They had a daughter Edith, a couple years older than my brother, and a son, Sonny, somewhere between Mickey and me in age. After Edith graduated from Vineland High School and enrolled in secretarial classes in Philadelphia, she often stayed with our family for extended periods of time. She was one of the lovely, older girls who treated me like a kid sister, initiating me with harmless, romantic snapshots of what to expect as a teenager and young woman. There was a large, old lilac bush outside the backdoor of their house, and one spring evening when I was about ten or eleven years old, Edith pruned a double cluster of the fragrant, pale-purplish blooms from the huge shrub and pinned it to my lank, brown hair. She told me that I was beautiful and that when I was older and wore flowers in my hair, a handsome man would fall in love with me. I blushed and thought of her brother but could not tell her of the crush I had on Sonny. They were all kind people, earthy and natural, hard-working farmers who valued learning for their offspring. Sonny's love of animals led him to veterinary medicine.

   Joe Lamnin, son of Sam's brother, was my age and became one of the beach friends during my teen years. Oddly, his high school graduation photo somehow was stored in a box of family photos—one of a handful that I did not jettison but did not carry with me when I left my parents' home. After our house fire in 1976, I found and rescued this slim packet of photos of nearly forgotten friends from long ago. Recently, I unearthed them again and searched for Joe on Facebook. His page showed no sign of recent activity, and he did not respond to my message calling out from the past. Blond and cute, he was one of the kindest, gentlest, yet fun boys I ever knew. I was pleased to extrapolate from his Facebook page that he had become a social worker with a career devoted to caring for people and championing disability rights. Joe had had polio as a young child and bore the lingering symptoms—in his case a partially disabled arm. He and his family seemed like people I could call friends now. Joe had inscribed his 1957 photo "Ruth, Lot's (sic) of luck in the future to a really swell girl. May our friendship last." The almost voyeuristic ability, bizarrely bestowed by technology, to garner a glimpse into the life of a friend not seen in nearly sixty years leaves me feeling that my reach is not quite long enough to connect again over time to either the old man he has become or the earnest, clean-cut, handsome young man he was. I want to call out to him, "Joe, it's me, your old friend Ruth."

The next landmark we come to on the corner of Evelyn Avenue and Gershal Avenue was the house of some Barish relatives, my grandmother's kin, but my memories of them are faded and bare no distinct traces of events recalled with lasting and special fondness. On the left side of the road were mostly fields with a few houses whose inhabitants I cannot remember, until across from my grandfather's was old Mr. Berkowitz (unrelated to the other family of the same name), the expert Kosher wine maker. He picked or bought his grapes in the late-summer harvest and made, then aged, the wine in his basement. Come early spring when we visited on a weekend, we got to taste the wine, sipping tiny tastes of the sweetish, fermented grape juice before he shared it with family and close friends for Passover Seders. He was a kind, elderly man who always seemed a bit remote and old-world. His daughter Betty, her husband Ben and their son David were close family friends whose home was in Philadelphia, and they remained close to my parents well into my parents' late-middle years. All these people were important in what seemed a self-contained, little Jewish community of full-time rural dwellers and the city-mouse cousins who came to stay in summer. Before Pop's on the right was a large chicken house where my brother worked part-time catching chickens in summers as he grew older. I cannot remember the name of those neighbors, the ones who ran the Kosher poultry business right next door, separated only by a field of corn or beans; they were merely neighbors, not close friends.

Farther down the road on the left were Jack and Hela Lehmann, and Laura their daughter, about nine years older than I and as sweet as a big sister to me. Laura often came and stayed at our house in Philadelphia after her high school graduation but before marriage and motherhood. The Lehmanns, like some others in the area, were Holocaust survivors. Hela possessed what seemed to me a most amazing skill. From her youth she had been trained as a glove maker, crafting kidskin into snug-fitting, fashionable, wrist-and elbow-length gloves that women wore in those bygone days. Combining machine and impeccable hand-stitching on the ivory or dyed fine leather, she made casual and formal gloves for every occasion any well-dressed woman in the 1940s and 50s might encounter. My mother and aunts owned several pairs, and I was thrilled when Hela, on one or two different years, made a special birthday pair for my small, busy-fingered hands. Jack Lehmann's brother and his family lived in Vineland; their daughter Judy was my age and a friend during our teen years. Hela, Jack and Laura's house was about a city block down Gershal Avenue on the left side, close enough for a mother and small girl to walk or for the generations to go together and enjoy a cup of tea or coffee and Hela's tasty coffee cake. Hela poured a glass of milk for the kids to have with their slice of home-made, old-world cakes which she made in the European mode, weighing by grams on a kitchen scale rather than measuring in cups and fractions thereof. My mother learned her recipes by measuring what Hela weighed and writing her version on 3" by 5" cards. I believe my mother also learned to make strudel, the paper-thin, fruit-filled pastry, with Hela.

Another two or three "blocks" beyond the Lehmanns' was Jesse's Bridge that spanned the Muddy Run. Poppy pointed out catfish swimming in the shallow, grassy water and sometimes caught them to show us what they looked like up close before releasing them back to the tributary of the Maurice River. Our family had ceased observing kashrut or kosher dietary restrictions long before I was born when Pop was diagnosed with ulcers. Not only did the medical world not have a clue about the cause of ulcers back then, they prescribed specific foods, like oysters, for their treatments. Pop also had gout, kidney problems, glaucoma and was quite deaf. Although he took to oysters with gusto and forever after treasured an oyster stew, we never ate catfish. Jewish dietary laws, besides prohibiting pork, rabbit, and deer, the consumption of meat and dairy together, and all shellfish, also proscribe fish without scales. Many Jews who do not observe kosher dietary practices, nevertheless, still retain personal, irrational taboos about certain foods. I knew such a young woman who regularly enjoyed ham sandwiches but would never put a single slice of cheese together with her deli ham, lettuce and bread. So, we never ate catfish, perhaps not because they were not kosher, but because the wild, river denizens have a stronger taste than their farmed cousins that are so popular in fish fries and the all-you-can-eat catfish restaurants of today and because the fierce appearing piscines did not look appealing.

The walks to Lehmanns and back, or beyond to Jesse's Bridge, indeed, up and down the entire length of Gershal Avenue or anywhere in the area, yielded an abundance of wildflowers from spring through autumn. The common, orange daylily or Hemerocallis and the blue flowers of chicory and Centaurea, or bachelor's buttons, were weeds that grew along the roadside in mid-summer next to wild carrot or Queen Anne's Lace. There was also yarrow, and distinguishing between it and the lacy umbels of the royal lady's lace challenged me in my earliest days as an enthusiast of wildflower walks. Toadflax, or chicken and eggs, the tiny yellow and white wildflower version of snapdragons, was one of my favorites. Very early in life I learned the chant of "He Loves Me, He Loves Me Not" as I pulled the white petals from the ox-eyed daisies. In addition to wild iris and wild bleeding heart, there were miniscule wild strawberry blooms that would ripen into half-bite-size morsels of strawberry sweetness come summer and the similar white and yellow, simple, pleasant flowers of black raspberry and blackberry vines. The sweetness of honeysuckle blooms delighted us in early summers. So many others that I may have known by sight then and learned to recognize, not just by sight but by usage as folk medicine over the rest of my life, grew in the environs. Pop harvested cattails from swampy areas to ignite, attempting to repel the notorious mosquitoes of southern New Jersey.

In tenth grade biology, our botany project required us to gather and identify a stated number of wildflowers in a scrapbook format. Certain that I could easily and joyfully accomplish the task and earn an A+ for this exciting assignment, I enlisted my mother to accompany me on spring-weekend walks in the country. I determined that we could visit Pop every weekend until I had more than the minimum number of flowers to dry, mount and identify by their vernacular and botanical names. It was lovely time spent with my mother whose encouragement led me to arrange my dried and pressed wildflowers so carefully and artistically. My disappointment with my grade of B- when the projects were returned was so profound that I could not keep back the tears as I walked up to my beloved teacher's desk at the end of class to inquire about my grade. The teacher was a kind and patient man and an inspiring and passionate educator. I could barely croak out the question when I asked him why I did not get an A, and he gently explained that because I had so artistically cut off the soil covered root and quite possibly the stalks and leaves as well, I had eliminated equally essential, albeit not the showiest, parts of the plant. How drastically I had chopped the flower from the plant I cannot exactly remember, but I was clearly more impressed by the blooms than by the whole systemic organism.

Encompassed in this experience were two hard lessons to learn, both of which I have long recognized as vitally important. The biological lesson taught me that the plant in its entirety, like any other organism or ecosystem, is essential for continued life. As a gardener who still finds the germination of seeds somewhat miraculous, I know that the tiny thread-like root sprouts along with the dicotyledon leaves, if given proper care, organic nutrients, and good soil and water, will develop strong roots that anchor the plant where it must grow. The second lesson is that outward, or flashy beauty must rely on a solid foundation; in terms of plant life, the flower, though necessary for reproduction of the plant, will not survive without the roots. In writing this memoir, I am acknowledging, indeed honoring, the roots made by my great grandparents and my grandparents and the legacy that they and their community bequeathed to me.

As a child, I was lucky enough to be included on the occasions when Poppy took my brother, and boy cousins if they were around, to his favorite haunts; a stop at the Stavitsky brothers' "cabin" (Jake, b. 1896; Benny, 1902; and Barney, 1903) was at the top of the itinerary. The Israel Hersh Levin and Elia Stavitsky families grew up next door to each other. My childhood memories do not distinguish things like a difference of ten to twenty years in men who wore the similar, non-descript garb of farmers in the late 1940s and early 1950s: blue denim dungarees (as jeans were called then), perhaps, although all I remember with certainty is khaki, olive drab, brown—neutral colors that would not show the dirt and that would allow them to walk through the woods and fields dressed in some degree of camouflage. They were first generation Americans, Jewish Huck Finns of Southern New Jersey, at home in the fields and woods and at the lakes and rivers of their childhoods. They read *Field and Stream* and other outdoor magazines, hung their unloaded rifles safely out of reach of children, and decorated the house with deer antlers. The Stavitsky brothers had a legendary collection of arrowheads. The visual memory of being in their house where arrowheads covered the walls came rushing back to me when, in an email, Jay Greenblatt related how his forty-year-old efforts to preserve their house—the last standing, original settler home—failed for a variety of reasons in the late 1980s.

Rich Brotman's 1982 documentary *First Chapter in a New Book* includes interviews with the last of the children born in Alliance in the true Golden Age. Seeing and hearing these old men who still looked strong and vital and whom I knew in my childhood makes me realize now that although they were halfway between my grandfather and my mother in age, the Stavitsky brothers were buddies with my grandfather, more like younger brothers or nephews. While slicing mushrooms for the day's main meal today, I had a "kitchen epiphany," a momentary insight born of simple, satisfying domestic work, the likes of which have largely built this memoir, that he must have loved his friends as the sons he, a father of three girls, never had. While my mother and aunts took care of raising me as a daughter, I am grateful to have received any exposure to the traditionally male aspects of rural life as embodied by John Levin, the only grandparent I ever knew, and his cohorts in this small community.

∾

Alternative Routes and Atypical Adventures

An alternate route for the last leg of the drive from Philadelphia to the country jagged southeast for a few miles at Malaga and took us to the little hamlet of Newfield. For a span of years, a most unusual shoe store existed there. Whether they also sold men's and boys' shoes or just women's and girls', I cannot remember, but I have visual and tactile memories of shoes that my mother and I got there. The store consisted of a couple of dark, cramped rooms filled with unsold shoes from various upscale retailers. Each style might be available in one or two sizes, perhaps only a single pair each; it was hit or miss, score big or leave empty-handed. When I was about ten, my mother, whose glamour days were pretty much over, could not resist a pair of sling-back, strappy, high-heeled sandals in soft leather of coral and turquoise that sold for a pittance. She had some costume jewelry—earrings and a bracelet—that perfectly matched those shoes. The shoes sat in their box in her closet for years, awaiting a dress that was worthy of being paired with those shoes and earrings. Perhaps, like the one pair of spike heels I ever bought, their beauty greatly exceeded their comfort. We bought me a pair of black suede hybrid-style shoes that I rejected at first. Like the pumps I was in search of but unable to find, they were the middle-school version of dress shoes for girls with their one and a half inch-high heels as big around as the heel part of a flat shoe. The black suede slipper front had a sling-back—a very grown up feature—and an inch-wide band that fit across the instep. They were different than the standard pumps with bows my girlfriends wore, but I very quickly grew to appreciate them and my mother's avantgarde fashion sense as they garnered compliments.

From the shoe store, we took the Delsea Drive and drove into Vineland on its northeast city limit. Sometimes we stopped at Zucca's Bakery but more often went there on return trips to the city for their freshly-baked Italian loaves and flaky, fruit-filled tarts. Vineland's main drag, Landis Avenue, was as wide as a boulevard—so wide that cars could park diagonally to the curb without impinging on traffic. If the driver was lucky, he (my father) could get in sync with the timed traffic lights and never have to stop at a red light. Once we passed through the business district of Vineland, which held its own delights like the chain confectionery shop, LOFT's Candies, with its distinctive sign of full, clean, bold capital letters for the name, the bar of the "F" striped like a candy cane, and a lower case, equally upright, curly, cursive spelling out "candies"—as delicious a promise as a swirl of chocolate—I perked up.

After the business district thinned into a mix of single houses, farmland and woods, landmarks appeared on both sides of the road. There was the duck farm with its smelly pond and hundreds of white ducks on the left side.

Also, on the left was the sign and turn for the Palace of Depression, the bizarre, terrifying, and alluring tourist attraction that appeared on all New Jersey maps during my childhood years. It was constructed and run by George Daynor who sold twenty-five cent tickets to tour his strange house-cum business built from recycled glass, bricks, concrete, crushed and rusted vehicle parts and general junk. The Palace of Depression was a place we visited on special occasions when friends came to the country with us for the first time, on pre-beach season excursions, or days when we could not go to the river or lake or visiting, days when every other diversion failed and we all wanted to be titillated by the terrors and thrills of the fairytale world come alive in the inventive hands of its mad entrepreneur. To a small girl, George Daynor's appearance was frightful enough, but his dwelling and his stories were stranger still. A ticket booth that looked like it came straight out of a sinister scene in a Disney movie stood apart from the main attraction. While researching The Palace of Depression in recent years, I learned that this was all that remained after a fire in the 1960s. A couple of men who grew up in Vineland and visited the site regularly in their childhoods have organized a team of volunteers and continue the extended process of restoring it to its former, funky glory.[1]

A photo of the ticket booth brought back so many memories of this bizarre and fascinating place. I remember George Daynor wearing, or displaying, a leather helmet and goggles, the kind worn so rakishly by early pilots in open-cockpit, prop planes. While the walls of glass bottles and concrete were attractive to a child who loved the play of light and shadow through transparent and translucent substances and on all surfaces, George's menacing accounts of the Jersey Devil—a biped, bat-like creature of the Piney Woods—and references to bodies found in the bogs terrified me. He never tempered his generous repetition of these baleful stories to children and their parents. I'm sure that many gleeful children who visited the Palace of Depression in the bright sunlight were haunted by the cautionary tales in their dreams at night. In March of 2019, friends from Little Rock who were visiting family at the New Jersey seashore made a tourist stop at the half-rebuilt Palace of Depression and emailed photographs; these pictures trigger memories that pale in comparison to the ones long stored in my visual cortex.

On the right, a mile or so before our turn on to Gershal Avenue, we crossed the Maurice River. In my memory, that section of the river was very narrow and was accessed right from the road. Cars or trucks parked on the shoulder of the road, and black families swam there. In those days, long, long before the term "white privilege" entered the lexicon, I always had an inner awareness that our community was lucky to have a place on the same river. I could not have articulated what I felt then, but I knew that just a few miles by road up river, we had a beach; a pavilion with a food concession, a juke box and pinball machines, a dance floor, and booths where men played cards; wooden picnic tables outside under the pines; our own private swimming hole; and the histories of forty-three families and their descendants engrained in the very sand where we played and sunbathed and in the very eddies of the shallow river where we swam.

Brotmanville had already become almost a ghost town. Prior to World War II, one hundred families from the privileged German-Jewish lifestyle realized the growing danger of Hitler and chose to leave their once luxurious lives while they could. They settled in the area and became kosher poultry farmers. After the war, 350 young families, mostly eastern European Holocaust survivors, found new homes in the several small, intentionally agrarian/small-scale industrial communities between the South Jersey towns of Vineland, Millville and Bridgeton. In the mid-1940s to 1970s the population swelled to about 12,000, as Jewish farming, due in large part to the kosher poultry farms in the area, enjoyed its heyday.[2]

Chapter 7

The Maurice River, 1943 to 1959

The Maurice (pronounced "Morris") is a short, lazy squiggle of a river in southern New Jersey. About fifty miles long, less than a third the length of the state, its source is in Pittsgrove Township at the confluence of Still Run and Scotland Run below Willow Grove Lake. Its watershed includes the famous Pine Barrens, a unique wetland region of New Jersey. It flows from its source south/southeast through Salem and Cumberland Counties, through the cities of Vineland and Millville, where it was dammed to make Union Lake and past which it is navigable. It makes a short, southwest shift as it empties into the Delaware Bay.[1] The sandy, scrub-oak woods of South Jersey and its tea-brown tributaries, where my mother's paternal grandparents and forty-two other founding settler families emigrated in 1882, are imprinted on my memory. The Maurice River flowed past the small family farms and later Kosher poultry farms established by the Jewish immigrants and was the social hub during the brief summers of my childhood and youth. The Alliance/Norma beach was about two miles north of my grandfather's little "retirement" five-acre farmette.

Tinted a reddish-brown, an orange-pekoe brew from the tannins of the timber in the area: scrub oak, pine and cedar, its hue was deeper and redder in my memory than reality proved it to be when on August 11 and 12, 2018, I returned to the Alliance Community for a picnic celebration of the 136th anniversary of the founding of the colony. Today the Maurice is a Wild and Scenic River, designated as such by Congress and President Bill Clinton in 1993 and saved from becoming the state-sanctioned site for waste reprocessing by the toxic waste industry. Part of it was badly polluted by Vineland chemical industries, resulting in the establishment of EPA Super Fund cleanup sites beginning in the 1980s. Although arsenic and lead are still present in the sandy, silty bottom of the river with on-going cleanup protocols, it is now proudly touted as a clean river with a diverse ecosystem of flora and fauna.[2]

In the days following my whirl-wind trip from the Arkansas Ozarks to the Alliance Community, I eagerly checked Facebook several times a day. I had just reunited with once-close second cousins I had not seen in nearly sixty years and had made new connections with much younger kindred spirits descended from the original settlers. When a cousin who had grown up in the community posted pictures of the river at the old Norma Beach, one of her friends commented that is was sad to see trash at the beach and the way the river had reverted to a state of wilderness. Although my childhood memories of that place are among the happiest, most significant memories of my life, I was able to embrace the changes while recognizing what remained to trigger those memories. In fact, keeping natural areas free from litter and trash is a privilege of affluence. The beach at the dead end of Eppinger Avenue is private land, possibly owned by absentee landlords. If the state of New Jersey or even Salem County owned it, there might be well-maintained trash receptacles. I know nothing of the tangled web of real estate and the legacies of family land trusts and why the owners do not maintain the beach where generations had recreated. As much as I would love to be able to relive some of those times, I was in awe of the wild, free-flowing river and the rare wildflowers along its banks.

While I was there with a new-found friend, a group of about ten kayakers floated along. An avid, recreational kayaker of The Buffalo National River in Northern Arkansas and other Ozark rivers, I pelted them with questions about floating the river. They were kind enough to answer as the current took them downstream. Now my dreams about The Maurice River include returning with my husband to float the mid-section of its tea-brown waters.

∾

Childhood Years

After a house fire burned our half-built house and meager possessions to the ground in 1977, less than a year after my husband and I married and began our long life together, I gathered a slim packet of black and white pictures of the first half of my life from the photographic stores of other family members. The fire is another story, but to me no longer compelling.

Nothing arouses memories of my childhood the way a 2¼- by 3-inch, cropped, monochromatic photo does. It was taken with the old family Kodak "Brownie" on August 7, 1943, three scant months after my second birthday. A box camera packed almost like a little trunk with reinforced, metal corners, a leather handle, an ingenious clasp, and a shutter-lens like a winking eye, it felt like a marvelous artifact even in my youth when it probably was state-of-the-art for family snapshots. I say monochromatic rather than black and white because the photo has a slight sepia tone, resulting most likely from chemical fading. This precious, tiny image is an icon for me, a talisman I have kept near my bed in recent years during trying times to evoke the protective powers I had always felt emanating from my brother. For years, from childhood well into my middle years, I had a recurring dream of submerging in the water at a tributary of the Maurice River within walking distance of my Pop's place, The Muddy Run. Although I was not a good swimmer—in fact quite afraid of the water at times while also loving it intensely—this dream was not about drowning but about entering another realm. It is a watery threshold into the twilight of personal sub-conscious and collective community memories. This tiny pictorial record of a day at the river still educes so much of my early childhood history and emblazons all in full, sunlit color though it was merely black and white.

The five people in the photo are in, on or behind an enormous truck tire inner tube, which hung in its relegated place in my grandfather's garage for years. Fat, sleek and reflective as a seal, it bears two red rubber patches affixed to stem leaks on the tube. Looking at the photo all these years later, I can feel the edges of these patches against my fingers and smell the dust and heated rubber scent as the tube, made untouchable by the burning sun, is flung into the water. It practically sizzles as its mass hits the river and eddies into the current. My brother Mickey, a strong nine-year old, stands in waist-deep water outside of the tube on the viewer's right, his olive skin dark-tanned, his short black hair a glistening, wet helmet. If we move into deeper water, he can hang on and stretch his feet out to float, but he is clearly standing and piloting us downstream.

In the lower center of the tube, her chin hidden inside its curve, is Toby Berkowitz whose hand, at an odd angle, grasps the outside of the tube and is the focal point of the arrangement of figures. Her light-brown hair, bleached honey-blonde by the sun, is pinned in a crown of braids and shines, and a slightly angled line through her features and fingertips connects her to me and then to my Aunt Jean. A more perfect composition based on the principle of the diagonal line could not have been arranged by photographer or painter. It is reminiscent of a harmonious grouping in a Renaissance painting. We all squint our left eyes into the sun although the right half of each face is cast in shadow. Toby, a ship's figurehead on the great black inner tube, leans eagerly into the adventure.

The two-year old child that is me appears skeptical of what lies downstream and holds back, but truthfully, I can remember nothing of that day and only read into the photograph the experiences and reflections of a few years hence. The raised eyebrows in a tanned face, a face haloed by wispy hair bleached white by the sun and pinned out of my eyes with a white-bow, are not those of a secure, confident two-year old, and maybe my later memories of fear of the water constitute a ghostly presence of dread next to me in the photo. Maybe it is just a quizzical expression, a child beginning to engage with the world around her and trying to figure out what was going on. No pictures of me at the river in my first summer, when I would have been only a few months old, or of my second, when it would have been reasonable to introduce a one-year old to the water, remain. It is odd how the faded, thinned eyebrows of my old age retain the shape of those strange diacritical marks on the face of a child mirrored in the aging woman.

I am held inside the tube behind Toby by a woman standing outside of it. Aunt Jean, my mother's sister, whose mouth is obscured by my head, just as my cheek is obscured by Toby's braided crown, is dressed in a sleeveless white top with a white bow in her black hair, giving her a slight Geisha-like appearance. To the viewer's left of Aunt Jean sits a boy who was her son and my cousin, Bruce, about three years older than I. All but a portion of his face is lost in the shadows. The memoirist at work leaves Bruce obscured by his mother's shadow for now and moves to the viewer's left of Bruce to a sliver of a young female body, a faceless distant relative or family friend.

153

The photo is cropped there, excising her from the picture for an obvious reason: I probably cut it to fit into my fat, photo-filled, teenage wallet. Some lingering sense memory also recalls from long ago a faint, shimmering shape on the water in front of the inner tube, the shadow of my mother snapping the picture, which was also cropped. My relationship with Mnemosyne is imperfect, and contemporary brain science has revealed so much about memory's transforming processes that all I can do is wonder and invent richly within self-delineated, narrow confines, if I choose that path. When writers ponder what compels them to tell the story, perhaps the need to stitch fragments into a whole again is salient. I would glue the figure of the young girl and my mother's shadow back into the picture if I could.

The background of the photo is a truer record of history. There, tiny figures of several women in skirted bathing suits stand in hip-deep water in front of a walking bridge, which connects the bank of the river that is a sandy beach—the sand having been brought in by truckloads when this 'resort' was built at some, for me, unknown time before the mid-1920s—to the wooded bank where the current was most swift and where "water moccasins" purportedly lurked. The lower parts of the bridge, most likely of cedar and other long-lasting woods with pine for the railing and cross pieces, are still intact. I only dimly remember the bridge in this state, possibly only from this photo, but I seem to have some faint muscle memory of walking across it in a state of anxiety in childhood. By the time I was a teenager, the bridge could not convey the walker but still had a few piers from which strong, agile kids—mostly boys flaunting their bravado and physical prowess—dove when the water was high enough. But here memory muddles again.

Behind the bridge, the view of the river narrows and the river bends to the viewer's left towards a place called Dopey Bill's, or just Dopey's for short. Legend had it that

Bill was a recluse, possibly someone with mental impairment or else a profound disdain for his fellow humans, who built a little shack amid the scrub-oak and stunted-pines on the banks of the Maurice River at some unrecorded date long before my time. Dopey's was also the proud bearer of the nickname "Bare Ass Beach" because generations went there for skinny dipping or what my parents' generation called "buck bathing."

There is another photo from that day. Uncropped and without minor characters in the background, it is of my beautiful mother and me at the river's edge. The composition is not complex and interesting and doesn't please my artistic eye so much as it comforts my inner child. Because it is a relic and rare, it is endowed with magic for me: Madonna and Girl Child at the river's edge. My mother's dark hair, also in a braided crown, and her black, tank-style swimsuit give her a glamorous look as she crouches in the ankle-deep water, an arm around my shoulder. I am dressed in only white panties, a skinny, tanned child with wild, sun-bleached hair and eyebrows askew.

∽

There was a sense of safety at the edge of the water on the manmade beach where kids played within reach of the adults. The curve of the river provided a gentle harbor for community and friendship where every kid could go to every mom within circles of family and friends to have their needs fulfilled, a scraped knee cleaned and band-aid applied, a sandwich or piece of fruit and a drink provided to a hungry and thirsty child, a layer of Coppertone applied. We played there at the simple pleasures of an idle day at the beach. Filling and refilling the empty glass soda bottles and taking pleasure in the glugging sounds the water made as it bubbled into the bottles until day's end when we returned them for the few cents of a deposit or got a pretzel in exchange. Building sand castles and digging the deepest holes possible or extracting the greyish-white clay from the bank to form mud pies and clay vessels or to apply as facials. The mining of clay and burying each other in the sand were the innocent pleasures of children who took everything for granted. Even in our early childhoods we knew we were part of a unique and special community whether we lived there or came from the city to visit grandparents who were still part of the farming community.

Somewhere around the age of seven, I made it clear to everyone that I was like Daddy—not a good swimmer. Mommy and Mickey were naturals. I must have always needed to check that they would be there for me in deep water, but the one time that I was not crying wolf to test them, my brother was talking to friends and being cavalier about my real cries for help. I had somehow stepped or floated into the current and was over my head and going downstream. I jumped up and down, waving my arms and screaming "help" until a thirty-some-year-old male we didn't know saw me and ran in, plucked me up and carried me to shore. By then, Mickey was cognizant of my needs and at full attention. I was unharmed but frightened enough to promise to stop crying wolf and not go in the water alone unless someone was really watching me. Mickey and I shared the blame and responsibility for that mishap. I dramatized it as much as possible to my parents without blaming my brother too much, but my parents acted wisely about it and we all committed to more caution until I could swim better and could be trusted to not go in the water alone.

Jewish-American children in the mid and late forties were privileged creatures. We were reminded again and again of the starving children in Europe when we balked at eating the good food put in front of us. We were loved and protected in ways that probably defied our parents' and our own abilities to understand and articulate. In our parents' minds there had to have been a sense of enormous blessing, especially after World War II when holocaust survivors found refuge in the community. We were the golden children who surely would have perished had we even been born in Europe. My generation's parents, themselves the grand children of the original forty-three families escaping Tsarist Pogroms, must have felt a blend of privilege and wonder, and some guilt, for not being able to help more relatives escape from the Nazi destruction of European Jewry.

Every July 4 the Pioneer Women of the Alliance Colony put on a picnic at the Eppinger Avenue Beach on the river, and my memories of these picnics remain the coins against which I have judged all other community-based social events. Only in Arkansas have I found a similar sense of community, a place where transplants, especially "back to the landers," who themselves are emigres from another world, have formed bonds as close and supportive as those of blood and family. My memories may in part be conflated with other picnics there or at Pop's because I remember worn wooden picnic tables under the shade of oak and fragrant pine trees laden with home-made kosher food: Hebrew National all beef hot dogs, hamburgers and Jersey corn grilled on charcoal fires, huge bowls of potato salad, coleslaw, macaroni salad, sliced Jersey tomatoes, watermelon slices, soda pop—but no ice cream. There was probably Jewish apple cake made with oil instead of butter and milk. We purchased tickets from fat rolls and redeemed them for plates of food. Perhaps the childhood picnics of my generation are at least partially as I remember them because they had changed to meet the food tastes of the times.

Helen A. Oxman's memories come from an earlier time and another menu. An article about the 1982 centennial cele-bration of The Alliance Colony in *The Philadelphia Inquirer* tells how Helen's Philadelphia family—not descendants of the

157

settlers—bought a clapboard, former general store as vacation house and, thus, entered the community life of Alliance/ Norma. Helen explains that the Pioneer Women's picnics, sponsored by the Kanot chapter of Vineland, included more traditional Jewish delicacies than my memories do.

> Stoves would be set upon the beach, to keep delicacies like stuffed cabbage, kasha, roast chicken and roast beef temptingly hot, while cloth-covered tables were heaped with potato salad, cole slaw, chopped liver and plates full of sour pickles and seeded rye bread. For a special treat, lukshen kugels (noodle pudding) in many varieties—sweet, fruit, plain, were offered as side dishes.[3]

Reading Helen's description jogs my sense memories and brings back visual snapshots, though more like tatters of once fine silk banners floating on bamboo poles in a misty mountain pass. I do have clear memories of being cautioned many times about running and rough physical play around the tables at the picnic. In those days I would not have eaten some of Helen's favorite comfort food, but now kasha is a staple of my diet—for its flavor and super food qualities.

∾

Teen Years

The summers of 1956, 57 and 58 were the last ones of which I spent significant periods of time in Norma, and even then, I came and went between Philadelphia and the country, jockeying with family for the maximum time there. Sometimes I took a girl friend from school. My second cousin Ronnie Levin lived about a mile and half down Gershal Avenue with her parents and pesky younger siblings, and I was used to walking there from Pop's to hang out on days when we were not going to the beach or to catch a ride to the beach. Their house was always a second home to me. There was always a car stuffed to capacity to give me a ride home after swimming at the Norma Beach on the Maurice River. When we wanted a change from the beach scene, we would drive past my Pop's to Jesse's Bridge from which we jumped into the water. Though there are noticeable gaps in the concrete rails in 2018, the bridge still soundly spans the water. The Muddy Run, where catfish and carp are still plentiful, is an underground spring/shallow creek emptying into the Maurice's west bank a few miles downstream from the Norma Beach.

On summer nights for two or three seasons, we went to pool parties and square dances at Hirsch's Hotel on Eppinger Avenue or to house parties on the weekends at various kids' homes in Vineland. Fitting into teen life in Norma and Vineland was easier and more natural for me than it was in Philadelphia because of knowing the river from infancy as I did and finding it a comfortable and intimate part of my life (aside from that one near drowning incident that caused anxiety for the rest of my eighth or ninth summer). I was enthralled by watching my brother, older cousins and family friends metamorphose into that most fascinating of all creatures in those days: American teenagers. World War II was over, Democracy victorious, we were in an economic boom with the bitterness of The Great Depression fading. For working and middle-class white people, it was a golden age. Oblivious to the Cold War, young people had a place in the sun, although the pressure to conform to the accepted norms, modes of thought and even fashions of the 1950s marked it as a very conventional period.

The pavilion at Alliance Beach had doors in each direction but north. The food concession was at the end to the right as one entered from the beach. Couples, gaggles of kids, and men and teen boys, taking breaks from their card games, ate their hotdogs and hamburgers, drank sodas or licked ice cream cones and popsicles. Two or three pin ball machines occupied a strategic location near the juke box in its out-of-the-way corner near the counter where we stood to order our food. Booths lined the walls most of the way back, leaving a longer than wide strip of space where we would dance on the wooden floor. I don't remember the music from my brother's teen years, but I remember watching him and his friends with fascination. By the time I was fifteen and sixteen and old enough to be at Pop's by myself or with my mother or one or my two aunts, the music was different. Elvis, The King, was on the top 40; Johnny Mathis' crooning of standard ballads with all the orchestration and reverb made me swoon. We danced to everything from teen-focused Pat Boone hits and Marty Robbin's "White Sport Coat and Pink Carnation" to Mickey and Sylvia's "Love Is Strange"—which I always liked, partly because my brother's name was Mickey and my mother's Sylvia and for its syncopated but bluesy rhythm. The Everly Brothers, Harry Belafonte and Sam Cooke shared space with the Del Vikings' "Come Go with Me" and Fats Domino's "I'm Walkin'." Bill Haley and Chuck Berry got us rocking. My favorite slow song, though, was "All in the Game" by Tommy Edwards. It was first released in 1951 with full orchestration and heavy strings. A later, more stripped down and soulful version came out in 1958. The 1951 recording must have made the charts and remained on the jukebox long before I hit my teen years because we danced to it before the 1958 version was produced.

The beach scene was a social paradise for any kid who could dance, and I was a great dancer, having studied ballet for almost ten years, and capable of any Terpsichorean mimicry, but dancing shared the hours at the beach with sunbathing and swimming. Maybe my dancing secured me a place to park my stuff on that large, welcoming, abstract quilt of a beach blanket. We danced in the late afternoons, during summer rain showers, and after lunch. I danced with all the boys and felt sought after. In Philadelphia, where dating and going steady were the mores, I only got dates for dances because there were enough boys with two left feet who asked me to be their dance-teacher date to last through high school years. They were never the boys I wanted to date.

In my fourteenth summer, I was clearly still a kid and a tomboy—at least when it came to climbing trees. For years with an assist from Pop or Dad or Mickey onto the lowest branch of the black walnut or mulberry tree, I could scramble midway into its canopy. Of average height, I was always underweight, wiry and sure-footed. A photograph of me hanging from that lowest branch, which by then I could reach with a jump and pull myself by my arms into a position with my legs wrapped around a stout branch, shows me as a kid. The same tanned face and limbs as the two-year old in the inner tube, the two-year old with skepticism drawn in her eyebrows and white wisps of hair. Gone was the sun-bleached lightness of the hair, but the tree climber was still a kid, albeit a confident, happy one in the fresh air of my grandfather's rural paradise. That summer was transitional. I began to have more of a sense of the world at large and what impact it had and would continue to have on me, whether I knew it or not. At fourteen, my cousin Ronnie and I relied on her mother or another adult—relative or neighbor, or occasionally even her first and my second cousin, Barry, who was two years older, to give us a lift to the beach. On rainy days, which seemed in the predominance that summer, and which would have pleased all the farmers, if not the visiting summer kids, we sat on her parent's screened porch with Joey Shreiber, a year younger than we, but like me a summer kid. His grandparents on both sides were early Alliance families, and his lovely mother remained friends with my mother all their lives. Ronnie and Joey were content playing gin rummy on those rainy days when I wanted to be at the beach. Out of embarrassment that I would surely lose, I scoffed at learning the game that they both already knew so well. I may have had a little ADHD as well, especially with pent-up energy being held back by the rain. On days when there was not much else to do, I learned and participated, all the while pining for the beach. I was more a watcher that summer than anything else, and I likely read all the books from the school summer reading list, and more. On the days I stayed at Pop's, I would wrap myself in the heavy canvas hammock to read.

∽

1956 and 1957

The photo of me at the beach in my new, end-of-the-year-sale, latex, one-piece swimsuit with gathering stitches on the shorts and cuffed bra-top shows a girl one year older—fifteen, looking more like a young woman, though still clearly a child in many ways. Alone on the beach with bent knees and arms stretched behind me to support my leaned-back weight in the glamor shot, I look optimistic, happy. In the next photograph, I might make faces and mug, but I was becoming a little more comfortable about what I saw when I looked in a mirror. Inside all was not so self-assured. I was popular at the beach, part of a group of happy teenagers. Everyone liked everyone else and kids danced with one another unless they were dating or going steady; then they danced only with each other. Sometimes girls jitterbugged together. But I was still completely inexperienced. The whole social scene and my brush with popularity made me a little giddy, and I was not prepared for the close encounter I had with a very popular boy that summer. To me that episode will always be "The Debacle on the Screen Porch."

The hierarchy of pre- and teenage kissing activities in the 1950s began with the innocuous game of Spin the Bottle. My peers and I had practiced that adolescent ritual starting in the sixth grade in full view of each other in Philadelphia neighborhood living rooms while parents cast watchful eyes from their kitchens. It was a game of chance, as arbitrary as a spinning roulette wheel, with petty stakes and payoffs of kissing each other chastely on the cheek or innocent lips. Bland stuff, pre-kissing practice. Mid-level was the game of Post Office, but aside from an image of the couple walking into another room, I remember nothing of the procedure. Since memory leaves me with no examples, I deduce that I did not get to play much Post Office, was neither often the chosen nor the chooser, yet was probably greatly relieved. So, it is not surprising that with dim recall of that intermediate rite, I knew zilch about the apex of the triumvirate: Making Out.

I liken my experiences of learning to kiss to that of learning my times tables. Frequent and prolonged bouts of childhood bronchitis sometimes kept me homebound for up to ten days, such a circumstance occurring during the multiplication unit. It took years for me to gain any easy skill with numbers above five; fortunately, calculators became readily available and saved the day. Unfortunately, no device exists to calculate the intricate combinations of the social and sexual explorations of young humans coming of age.

Early in the summer of 1956 I met a boy from Vineland, first at the beach and then at a party. I'll call him Mark, and his friend, the one with a car, Steve. I had brought a girlfriend from Philadelphia to my grandfather's for a long weekend, a relatively new friend who had moved from Punxatawny, Pennsylvania. I liked her smaller town ways and knew that she would love spending time in the country. After the party, Steve and Mark drove Sally and me back to Pop's. Mark was cute and seemed to like me, so naturally I felt flattered. There was no scintillating, electrical static between us, but it felt like high time to get my feet wet with making out. That the Post Office lacunae in my kissing repertoire might be an obstacle failed to occur to me, kind of like enrolling in an advanced calculus class when all you could do was add and subtract.

We sat in nearly total darkness on the multi-purpose screened porch, and he pulled me onto his lap on the rocking chair where he sat. Luckily, the cots were not set up for sleeping as they often had been in my younger years when the house was full of family. I was ready for some exploration within well-defined limits. We talked in whispers, with Sally and Steve on the other chair. Mark and I played with each other's hair, and suddenly he was kissing me. I puckered my lips as I had seen movie stars do on the silver screen but was totally bewildered and grossed out when his open lips and tongue probed my closed mouth. In that instant I was plunged into a profound teenage angst and longed only for death—far more ordinary an experience than any girl undergoing the humiliation at the time could ever know.

But not every girl was the only granddaughter of John H. Levin, crossing guard at the Norma Elementary School and ruling patriarch of a three-daughter family. In the few seconds that followed my initial embarrassment, the living room and porch flooded with light, and there stood Pop in his boxer shorts and T-shirt, wildly pointing the beam of his flashlight, demanding to know what was going on, sending those boys flying—though out which door I cannot remember. I cannot remember his words; there I was caught in a pas de deux of mortification trying to assay which of the fatal steps was the lowest moment of my teenaged life, while also feeling that I had been reprieved. Sally and I fled to the second bedroom across the hall from Pop's and locked the door, while the boys ran out and jumped into Steve's car. I was certain that the sound of the car wheels spinning gravel in the driveway could be heard by neighbors up and down the country road, broadcasting the news loud and clear.

The next day, Pop only asked who the boys were and never said another word after learning they were from Vineland and not from one of the several other tiny Jewish farming communities in the area. Aware of the limits of his jurisdiction and influence, he nevertheless knew that he had driven his point home. No tatters of any ensuing situation remain, but I can still feel the heat rising in my cheeks. No whispering and pointing followed the next day at the beach. I probably told the part about Pop bursting in to my cousin and a girlfriend or two and hoped that that story would trump any 'kiss and tell' gory tales by Mark, revealing that I didn't know how to kiss. Soon after, he left for his job as a waiter at summer camp in Pennsylvania or New York state, and we never had another kissing encounter, though I remember that we wrote a couple of letters. He was gentlemanly on the few occasions we did meet, and the story assumed its natural status and finally fizzled out. My grandfather's reputation as a staunch protector of children spread its powerful wings over me.

In truth, events of the national scene that crept into our protected life were of a greater magnitude when it came to loss of innocence. I could not have articulated it then, but the horrific murder of Emmett Till the previous summer had affected me deeply. He and I were the same age, fourteen, when he was murdered, and I remember thinking that he was just a kid, like me. Throughout my high school years and beyond, the Civil Rights Movement left me more changed than my anxieties about kissing, which plagued me until they didn't. That story was already anticlimactic when it happened although I heightened its minor relevance when my anxieties about the powerful and bewildering attraction to the opposite sex gripped me.

By 1956, our country cousins had brought all their high school friends into the circle. Of all the nice and cute boys at the Norma beach in the summers there was one boy from Vineland who had no descent from Alliance roots nor connection with families we knew personally. My mother's first cousin on her mother's side and her husband socialized with his parents, but that was the extent of any immediate connection between our families. Even at sixteen and seventeen he had a Humphrey Bogart kind of cool and a James Dean sense of brooding. Whenever he was around for that summer and thereafter, my sub-conscious emotional traffic light went crazy: the red, green and yellow lights flashing out of sync, all at once, in short and long bursts, signaling conflicting messages of go-go-Go, STOP, proceed with caution, all the while illuminated by lightning bolts of coruscating excitement. Tall and thin, with long-sinewed muscles, he had crewcut brown hair and dark brown eyes. He was not cute: possessed of no dimples, no adorable smile, no silky, bouncing curls nor long, dreamy eyelashes. Instead his look simply stated that he knew things I did not know; it went well with a sardonic intellect and sheaf of clever lines thrown as bait to this chaste girl from Philadelphia. He was smart, iconoclastic, talented, from a family with liberal political views, and I was idealistic, an avid reader, and impressed by such a boy even while clinging to the romantic, musical tropes sung by the balladeer Johnny Mathis, whose music just sent me. The boy abraded some of my more romantic notions in our earnest discussions with his love of a rawer, more authentic music, and what I recognized then as a superior intellect. Sixty-one years later I see us in black and white, like the photos from that era, on the big Hollywood screen of memory, playing out a flirtation, trading quips a la Bogart and Bacall or the other famous B and B of romantic banter, Shakespeare's witty lovers, Beatrice and Benedict.

I first became aware of him, I think, towards the end of the summer when I was fifteen. We were not introduced at that party then but met on the beach early the next summer. I had turned sixteen in May and so had The Boy. My cousin and several girlfriends from Vineland came to Philly in May for the Sweet Sixteen birthday party my parents made for me at our house. Nothing fancy at a social hall, hotel or night club, but the food prepared by my father was far more delicious and appreciated than any catered food at any venue, anywhere. That was my summer to have fun, my summer of the mad crush and dizzying flirtation.

The teen section of the beach occupied the center, with nary a pine or oak tree for shade; we staked out the prime location of the sun worshipper's universe where we basted ourselves with tanning lotion or baby oil and iodine on a large, abstract, three-dimensional quilt made of adjacent beach blankets. The "quilt" had an irregular, but more-or-less, vertical delineation of girls on one side and boys on the other with an equally irregular border of couples. We were all within a couple years of each other and exploring our social relationships and emerging sexuality. Some of the kids were very fast, boys and girls alike. Some, like me, were innocent beyond belief, late bloomers, awkward with their inexperience and ill at ease with the surging hormones for which they didn't quite feel ready. Most were probably healthily in the middle, in varying stages of early exploration.

He, let's just call him The Boy, was already sexually experienced. It was just a known fact although I don't recall chatter and gossip about it. He walked up to where I was on one of the squares of the beach blanket quilt with my cousin and girlfriends, bent down, kissed me on the lips, and with a straight face asked, "What did you say your name was?" I laugh now as I tap out the words, feeling a tenderness to this boy who was in love with his own cool and the girl trembling with excitement and fear. I'm sure I marshalled my best Bacall sophistication and replied, "I didn't, but it's Ruthie," casually adding even though I was fully aware of his name, his age, and sundry details of his life, "And yours?" My cousins all knew me as Ruth Ann, but I was trying for a name with a greater high school persona and cute cachet in the era of Sandra Dee. With that kiss, he fired the first salvo of a three summer-long bumbling, beach flirtation filled with swimming, dancing, conversation and awkward physical incidents when he tried to kiss me. Each time that happened my yearning, but fearful and inept ego held on to my purity, like a rookie in an Olympic luge competition, while my young body screamed "lean into the curve" and let him.

When we felt sufficiently baked by the sun, we went into the water. We only went straight in from the beach for a quick dip or when boys grabbed girls and threw them in. The water was shallow, and in its natural curve, which still exists today, sunken leaves blown by the wind decayed. When teenaged boys in the 1950s behaved like brigands and threw girls in the river or dragged them into the muck, we screamed in mock protest and giddy delight. For almost forty-five years I have been an organic gardener living in the woods; I have long known that muck can be benign, a ripe compost pile after the rain, gentle silt for feet to sink arthritic toes into and wiggle in the curative texture. In my childhood and teen years, I dreaded the touch of muck. We were so socialized to find natural processes like decay repugnant. Adjacent to the muck pit a curve in the bank of the river served as the kiddy "pool," a shallow, flattish area crowded with mothers in lawn chairs, kids playing on the sandy beach and in the water. The kids were younger, some of the mothers were younger, but the social activities were the same as they had been a scant ten years before. The mothers commanded full view and dispensed attention and care to whichever child needed it as the kids dug clay and filled soda bottles. Teenagers steered well clear of the kiddy beach unless the boys felt the urge to drag the girls into the muck.

When we chose to swim farther from the beach area where everyone congregated, we usually walked up to the very small inlet where the water sometimes rose well above our knees and connected us to a spit of land and the vast wooded area beyond us, upstream where the river narrowed and curved, where the bridge that connected the banks still stood in my early childhood. In my teens, only the piers and parts remained, and when the water was high enough, very athletic kids could shinny up, balance, and then jump or dive off. I may have jumped once or twice with assistance but never dove. The boys were boisterous. Upriver where the water was wider, deeper and yet safe enough, they'd swim under the girls, surface upright with us on their shoulders and then flip us off backwards into the water. Of course, if you were interested in the play you were in the water, not off on your own at a picnic table with a book, dressed in shorts and a shirt or blouse, or on your own blanket at the fringes.

In 2018, in a completely different era, in the first year of the Me-Too Movement, in a time when, although still sometimes under threat, the LGBTQ community finds a more inclusive attitude and acceptance of diversity of sexual preference and gender identity, I reflect upon the conformity of the 1950s. My cousin Bruce, who was about four years older than I, was gay. The age difference alone would have had us in different areas of the beach blanket configuration and would have had him more absent from the scene for summer employment in Philadelphia before I had entered the magical period of fifteen to seventeen. I was too absorbed in the intrigues of my crush on The Boy and the protocol of teen socializing to remember a lot about Bruce's coming of age in that place. I do remember a girl about his age, either from Alliance or Vineland, who was cute and peppy but had a reputation as a fast girl. For some reason, probably an effort to prove the heterosexual masculinity he knew in a deep part of himself would never be his, Bruce asked her out a couple of times and got stood up each time. I knew by the time I was a senior in high school that Bruce was gay although that is not the language of the time. I don't remember a clear mental construct I had of that reality. I could identify some men as homosexual but did not find them or their mannerisms in any way repugnant. I liked them for their artistic and theatrical interests. Nevertheless, as a teenage girl and well beyond, I sometimes fell into the trap of stereotyping. Although we were close as kids, Bruce sometimes bullied me—mostly in the summers, especially if our other male cousin was around. The cousin my age seemed to me an athletic, alpha male even in childhood. When we had too much unstructured time together in the country, they were tempted to join forces and dump me from the hammock, or twist their hands on my arms to give me "rope" burns, or torment me in other ways. While my brother was with us, they behaved well. I forgave Bruce for the moments of bullying when we were very young and asked him to escort me to my senior prom for two reasons: he was a great dancer, and going with any other boy but The Boy would have left me completely miserable.

I recall mostly respect of other kids' limits, and kindness—a gentle, encouraging behavior devoid of any life endangering activity. There was no drinking at the river although possibly fathers had something on the side, and maybe some of the older or more precocious kids did. The Alliance Community's roots did not include alcohol as a tool for releasing social inhibitions; for Jews wine is part of liturgy and observance. I did not even know that Jews could have alcohol addictions or be habitual drunks until I was an adult. Fortunately, teenage drinking was never an issue, another aspect in which I felt sheltered and safe then. I was naïve, and may have been oblivious of bullying, cruelty, abuse of alcohol, violent behavior and the darker aspects of teen life of even the innocent 1950s. I don't recall auto accidents that killed kids, but life in the flat, South Jersey Jewish farmland of the 1950s is a far cry from the dark side of contemporary American culture and the teen drinking, drug use and fatal accidents that seem so tragically prevalent in the Arkansas Ozarks. So, it seemed safe, comfortable, not too challenging, a good jumping off place. But for all this cushioned platform from which we would launch into the world, the time in which we grew up was also stifling in its conformity. There was a Beat Scene out there in the world, not in Alliance, Norma or Vineland, New Jersey, nor in my row-housed working-class neighborhood in Philadelphia, but somewhere, when I became ready to find it. My clothes, my hairstyle, my taste in music were conventional, my politics idealistic but unformed; I needed to know how to fit in before committing to breaking out in any whole-hearted rebellion. When it comes down to it, we were living a bourgeois existence.

∾

Buck Bathing Through the Generations

Elizabeth Rudnick, the grandmother of several of my second cousins, reminisced in 1932 about the post-berry picking reward of a refreshing swim at the Maurice River in her youth, replete with "relaxation, laughter, revelry and games," but without bathing clothes. Elizabeth's anecdote demonstrates an ease and comfort with her own body which I find delightfully surprising for the time. "Bathing suits, of course, were out of the question, and there would naturally result all sorts of embarrassing predicaments, when the more daring boys would disregard the rights of 'possession' and usurp the water occupied by the weaker sex." Or, if the scene were laid in that part of the river . . . "where the seclusion gained by separate bathing spots for men and women would often be shattered by quietly swimming, cunningly contriving boys, who, when least expected would astound and frighten us by their sudden, hilarious appearance."[4]

I knew that my mother's generation went skinny dipping, or what they called buck bathing, in an entirely different social setting than that of the previous generation of hard-working, immigrant farmers. My parents and their relatives and friends were newlyweds who enjoyed socializing at the river at night, as well as during the day. I imagine them around a bonfire and swimming without the confinement of their bathing suits in the light of the full moon.

By the time my cousins, friends and I were skinny dipping, the rules of the game were internalized, understood by all the participants. On weekends, holidays, and during the week when enough of us were present, and no one had to watch younger siblings, we would know by some signal in the hottest part of the day that it was time to go through the woods to Dopey's. Traditionally, the girls went first, single file or in twos where the path widened. It was probably no more than a thousand steps into the woods where the bank was wide and clear enough to sit and to take off our modest swimsuits, hang them on tree limbs and slip into the water. There were girls, six and eight at a time, giggling, talking about the boys we liked, typical teenage chatter. After twenty to thirty minutes of freedom from the confines of latex, nylon or cotton one-piece bathing suits, we heard the boys coming through the wooded path and scrambled and wriggled to pull our suits back onto our wet bodies, a trick on solid ground, an acrobatic challenge in the water. Degrees of daring existed, but my girlfriends and I were not huge risk takers. We were all back in our squishy suits and swimming or floating in inner tubes downstream through the wooded banks by the time the boys made their noisy arrival. By then we were chilled from the narrower, swifter section of the river where overhanging tree branches shaded the water. We were eager to bake our bodies again in the waning hours of afternoon sunshine.

Skinny Dipping at Dopey's

The boys called Dopey Bill's bare-ass beach
though only the trailhead touched the edge
of beach made of dump-truckloads of hauled
yellowish-tan sand. Maybe my mother or
grandfather knew when it was carved from
the pine and oak woods or when the
pavilion was built. I only know when it existed.
Nobody then in my childhood and teen years
remembered who Bill was and what affliction
gave him the cruel name. Rural consciousness
was made of legend and myth: The Jersey Devil
and poor Bill, who may have been slow or mute,
crazed or outlawed. He must have lived by his wits,
fishing, hunting, trapping and maybe a moiety
of kindness by charitable ones. Bill was long
gone even when my parents and their friends,
newlyweds, skinny dipped at his swimming hole.
Protocol, an unspoken signal at a certain time of day,
moved the girls to the path at the edge of the yellowish-tan
beach to walk single file through the woods to Dopey's.
There we wriggled out of our latex or cotton one-piece
bathing suits, left them on the bank and slipped into the
tea-brown water. Quiet girltalk and giggles flowed like the current
until the would be-brigands approached through the woods,
trumpeting their presence. Then the struggle back into suits,
pulling damp second skins onto wet bodies like stuffing
pre-formed sausage into slippery limp casings.
Laughter and thrashing male limbs rang in the woods
as girls swam downstream and around the bend
when bare-assed boys cannonballed into the water.

Jesse's Bridge was about a quarter mile past my grandfather's house. There were times that our idyllic afternoons at the Norma Beach became too commonplace and we needed a diversion from the mundane pleasures of sun, water, dancing to rock and roll, and flirting. Unlike rural America in the 21st century, that day and age did not find most kids in possession of their own cars, even less so of pickup trucks. Someone with parents or grandparents still on the farm, possibly my cousin Barry, would sometimes be able to take the farm truck to the river.

We piled into someone's car or the rare pickup truck and drove the mile and a half from the dead end of Eppinger Avenue to Jesse's Bridge and pulled off the road to park. Then we climbed up on the westside of the bridge and jumped like lemmings into the relatively shallow waters of the Muddy Run. The Boy was usually irritated with me at the end of night-time social events when we were together in his friend's car for my ride home; however, when it came to jumping off Jesse's Bridge, he was as gentle and encouraging to me as my own brother would have been. Although it was not a large, deep pool with a strong current, it was the perfect depth in one spot to make jumping safe from injury and swift-water drowning. The first time we went to Jesse's Bridge together, I explained my twin fears: height and deep water. He promised to hold my hand and stay with me from the moment of balancing atop the bridge railing—the last chance for backing out before our synchronized jump—to swimming back to the river bank. And he did, faithfully, dependably, kindly until I grew confident enough to do it alone. Even then, we held hands for the rest of the summer when we jumped together from the bridge.

The bizarre neurosis I had about letting any boy kiss me was ridiculous, but he and I were physically close much of the time we were in each other's company. In my old age, even though I have never had children of my own, I understand and feel tenderness towards both of those awkward teenagers. When we danced to one of the ballads, our bodies heated by hormones but with still wet swimsuits, or to a fast-beat tune when we jumped, gyrated and twisted, aware of the changes that had taken place in the last few years, we were simply experiencing a natural sexual awakening. Margaret Mead or other cultural anthropologists interested in courtship practices would have had a rich vein to mine. When, as a group, we waded into waist and shoulder deep water to cool off, the boys continued upstream and, like submarines, stealthily swam underwater to where the girls talked, then surfaced between our thighs lifting us onto their shoulders, only to flip us backwards into the water.

Recently my husband and I watched *Mustangs*, a beautiful, powerful movie by Deniz Gamze Erguven, the feminist, Turkish director. In a contemporary Turkish village, five orphaned, lively, intelligent sisters are ruled with iron fists by an uncle, grandmother and aunts who tighten the stranglehold as they arrange marriages for each of the girls in turn. The first scene in the movie shows the girls and their friends— other girls and boys—wading fully dressed in their school uniforms into the surf on the last day of school. The boys hoist the girls onto their shoulders, and the girls wrestle with each other to see who can knock whom off the boys' shoulders. The tragedy of the film is the complete repression of the natural instincts of healthy young women. By the end I was crying, mourning for the joyous freedom of the young people that had been subverted and stifled in the name of rigid social norms, religion and the reputation of the family. I felt so fortunate that my own youth was as unfettered as it was, that I was loved and trusted and growing up in a culture that accepted youthful, healthy, sexual play in full view of an accepting society, even though I had not yet found my own comfort zone.

Our night-time relationship was less easy, more fraught with sexual tension. One night in July of 1957, our sixteenth summer, there was a dance at Hirsch's Hotel. The Boy was there, and we danced together around the pool. We walked off at some point, possibly so he could smoke a cigarette (although I cannot remember for sure that he smoked) and certainly for a bit of privacy, but were followed by a girl about eight years old. She was the youngest child and only daughter of a family—friends of my family—who had one of the several small grocery stores within an approximate four square-mile area. The father of the family had died a year or two prior to this summer. He was relatively young when he had a fatal heart attack and left, in addition to his wife and daughter, three sons: one a couple years older than I, one my age, and the younger about ten. The younger boy and his cute, little, blonde sister were ebullient, perhaps even wild, but sweet and affectionate. I loved having younger kids tag after me and, in the city, was the benign Pied Piper for all the younger kids on my block, leading them to the neighborhood playground a few blocks away and supervising their play for a group baby-sitting fee. This little girl probably saw me as the older sister she wished she had just as I had always wished for a sister of my own, so our relationship suited her purpose and mine. We were such a small community and kids knew their siblings' friends. We were like a village of children and teenagers well known to each other, and The Boy, who was friendly with her older brothers, liked her too.

However, her presence with us, on the porch swings or benches scattered in the park-like grounds of the hotel, did not suit his purpose then. He wanted to make out. Still feeling confused and awkward about my debacle with another boy the previous summer, when Pop burst into his darkened screen porch just as I was being officially and unpleasantly kissed for the first time, I kept inventing diversions. The company of my little friend, and then soon enough her energetic, slightly older brother, was one he could not accuse me of throwing up as an obstacle to his desires. Within a year after the father's death, the family moved to California where they had relatives, and I missed my little friend.

That evening when she and her brother followed us around, making us laugh at their antics, I was both delighted and impatient. I wanted to be with The Boy, feebly fending off his efforts at kissing me in the dark woodland clearings of the resort's grounds. The young siblings attached to us like limpets, so we wandered back to the poolside where records played that night. I was wearing white pants and a nice blouse or jersey, careful to go into the bathroom and check on my sanitary napkin. When the dance ended, the boy from Vineland who had a car and the other kids from there and I piled into the car for the drive to my grandfather's. This was before seatbelts and the maturity that they brought, but it was a short ride and there had never been cautionary tales of accidents on Gershal Avenue. The backseat was full to overflowing, and I sat on The Boy's lap, not for the first time, in the packed vehicle.

Earlier, either while we danced to a slow tune as we watched steam rising from the warm water of the pool into the cooler night air or when we walked off into the grounds, he initiated a conversation that was awkward for both of us. He was seldom awkward or at a loss for words, this boy who seemed so glib to me, smart, quick, confident—even brash—and gentle by turn. He said he had something to ask me, but stammered to find a smooth way to do it. I told him to just ask. Then he told me that one of our friends had said that I liked him. It was my turn to splutter and stammer while he asked, point blank, whether I did. My protests that I liked everyone dissolved and I admitted that, indeed, I did. I deflected his attempts at kissing me even while I relaxed into the embrace of his encircling arms. He accepted my shyness about kissing in the backseat with other kids crowded next to us, but when we pulled into my grandfather's driveway, he got out of the car and walked with me the few steps to the back door of the bungalow, out of sight of the kids in the car. By now I felt the telltale sensation that plagued every teenage girl and young woman—excessive menstrual blood flow—and was in a panic about getting into the house. Since there was no question of his coming into my grandfather's house, he tried for a big kiss as I successfully opened the door with the old iron key and slipped into the mudroom. I said goodnight and told him I had enjoyed the evening, but he just replied with rancor that if I did like him, I sure had a strange way of showing it. Once I got into the house and saw the blood stain on my white pedal pushers, I fell into a state of embarrassment and lay awake half the night worrying that the blood had also stained his khakis while I sat on his lap in the car.

Sometimes at Hirsch's that summer we swam in the pool at night. The colored lights around the pool tinting the wisps of steam that rose from the sun-warmed water into the cooler night air made an inviting scrim of romantic physicality. Another memorable evening at Hirsch's Hotel occurred that summer. A square dance was scheduled, and I had the perfect outfit for square dancing: red bandana-print cotton pants and a white, ruffled, tuxedo-style shirt. At some of the parties we swam and danced but usually not when a band and caller were present for square dancing. My skill as a dancer turned me into a different girl—poised and buoyant, a free spirit, unconcerned with success in winning his affection or anything else but executing the calls, skipping almost off the ground as we promenaded and do-si-doed, changed partners and danced the night away.

Again, the perennial driver chauffeured everyone home, starting with the few kids who lived along Gershal Avenue or one of the short roads intersecting it and then driving back to Vineland. I don't remember the back-seat tussle that night. He never forced himself upon me, so "tussle" may be misleading. What he did was endeavor to hold me in his arms and kiss me, and most of the time in those circumstances he got half way there. Unlike the boy "Mark" from the previous summer, The Boy did not French kiss me as an out-of-the-gate gambit in what Mark must have assumed would become a hot make-out session. Instead, as I chatted and asked him questions about his many extra-curricular activities in high school, what he'd read recently, what his after graduation ambitions were, he just kept trying to land his lips on mine and keep them there for more than a split second. I did not actually relent, but that night I softened my resistance. When we got to my grandfather's driveway, he walked me to the back door and tried again for a serious, lingering kiss. Meanwhile I, worried that my Pop's antennae would uncannily sense that we were embracing at the doorstep and come crashing through the mudroom in his skivvies with his faithful, Eveready-battery-powered flashlight, began fidgeting with the key. This was the same key that had temperamentally jammed now and then, and that summer was frequently in malfunction mode. I told him I had to get in, that we did not want to wake my grandfather. Gentleman that he had been trained to be, he took the key from my hands and turned it in the lock. Nothing! I tried, he tried. His friend in the car started flashing the lights on and off, signaling impatience. The Kiss, the one we really were so close to having, was suddenly a non-event. He tried the key again; I tried the key again. Fearful that the key would break in the lock, I resorted to a maneuver that I had perfected. I asked him to open the window, which I had left cracked for just that purpose. He performed my request with curiosity. Then when I asked him to boost me up to the window, his frustration and annoyance melted into surprised laughter as I vaulted through the open window into the mudroom, turned around and waved goodnight.

Thus proceeded our "relationship" whenever I was in the country, beginning when I showed up at the beach for a weekend with a sibling-type kiss of greeting and ending with gauche frustration for both of us when we were together at night. By now, I was able to serve a function like that of my mother or aunts, as helper to Pop, while enjoying the social life at the river and parties. My grandfather's health was more compromised than it had been in my childhood, and he was spending more time in the city at our house or with one aunt. I was able to do light cooking and cleaning, helping with laundry, writing letters and making calls for him. His hearing aids were far from the more advanced models of today, and he had trouble on the phone. Even so I was not in residence the entire summer but managed to carve out chunks of time beyond the weekends. The Boy and I picked up at the Norma Beach on the Maurice River throughout the summer, almost replaying the exact dialogue of the preceding scene. The settings were the same: swimming, sunbathing and dancing at the river, bridge jumping at Jesse's Bridge, pool parties with records for dancing, and square dancing with live musicians and a caller at Hirschs' Hotel, or house parties in town. The sixteen-year old girl that I was would have been happy with that scenario forever, but it could not last.

Soon after school started that fall, my cousin informed me that The Boy had a girlfriend. There were a few weekends during the school year when I visited girlfriends who lived in Vineland, and we went to parties. I remember a few parties he attended with no girlfriend on his arm, when we danced a bit and talked. One day when my school had a couple in-service days for teachers, my parents took advantage of my "vacation," and we went to see Pop to help him with some official business or health-related issues. We had arranged through my cousin and her school principal for me to spend Friday with her at Vineland High while my parents and Pop tended to his needs. In the physical education class, a

180

mixed volleyball game was in process, and both he and his girlfriend, whom my cousin had introduced me to in one of their common classes, were playing. She was friendly and nice. He greeted me but minus the warmth of our summer meetings at the beach. I was not a good volleyball player and was a little worried about getting hit in the face with the ball and having my eyeglasses broken, which had happened during a gym class game at my own high school. We were on opposite teams, and he was serving. BAM! Suddenly the ball he served hit me in the chest, knocking the wind out of me. I faked a quick, slick recovery, but the emotional pain, more than the physical sensation, was sharp, deep and cruel.

By my seventeenth summer, he came to the beach less frequently because he had a part-time job. His girlfriend was never at the beach, and I don't remember her at parties. We danced together still but not as we had the previous summer. He was not at parties as frequently, and it seemed that the parties at Hirsch's Hotel were also a less regular treat. I got other rides back to Pop's after night-time events and stuck around the house more, reading my way through my English class' summer reading list and extra books. *The Autobiography of Benvenuto Cellini*, which fascinated me with its vivid descriptions of luxurious and intrigue-ridden renaissance Italy, was one I particularly remember from that time. The attitude of ambiguous belligerence the boy had developed towards me was not a constant, but it flared now and then, and because there had never really developed a relationship with definable qualities between us, I could not ask him what was wrong. I got a glimmer of an idea at the beach one day that summer.

From the center of the beach where the mass of kids congregated, I saw him talking to two girls who had spread their beach blanket off to the side. They were not known to anyone, but as I peered at them from my vantage point, I recognized them. They were friends from my high school in

Philadelphia; we were classmates together in algebra class. So, it was completely natural for me to walk over and say hi to them. The Boy, as I approached their blanket, asked what I was doing, why I was following him. His response angered me, and I told him not to flatter himself, that I had come over to greet my friends from school. I did not know that Joan and Bonnie had connections with Norma and was happy to see them, eager to introduce them, and to make plans for that night that would include them. They had come to the South Jersey area resort area because Bonnie's parents, who were Holocaust survivors, knew people who had moved there after the war and knew that it was a nice Jewish resort in the woods and farmlands. I still liked The Boy and felt the same powerful attraction, and given the chance, I probably would have been ready for more physical intimacy that summer, but suddenly there was a side of him that I did not like.

It was not that easy to shed my attachment to this powerful crush I had on him, and the feelings persisted for another year even though he gave me little enough upon which to pin my hopes. I sent him anonymous, sarcastic Valentine cards to vent my emotions. The last time I saw him in Norma was the summer of 1959 at the pavilion at the beach. I had graduated from high school in January of 1959 and had a job as a file clerk in a center-city insurance company until September when I matriculated at Temple University's Teachers College. For years the Philadelphia school system had allowed students to begin school not just in September but also in the spring semester in January. They eventually phased this out, and my class was probably one of the last ones. He graduated in June of 1959.

Pop's health was getting more and more unpredictable. His property would soon be sold to pay for his medical care, and what I saw as our family's true home in Norma, New Jersey, would soon be lost to us. It was the last time I was there as an independent teenager. Fewer friends were there, having graduated and gotten summer jobs before going away to college. Nothing was the same. I was there with a cousin from Philadelphia to attend our Levin Cousins' Club picnic. The Boy seemed pleased to see me. We talked for a while, danced there one last time, and then my cousin and I left.

I would never again spend time there in the same way that I had. Everything I knew and loved and identified as the framework of my life was coming to an end. My childhood summers in a twentieth-century, post-war, Jewish Acadia were over. A world of adult responsibilities awaited me. My summer romance had budded, bloomed, and withered but never bore fruit, leaving me no better prepared for the sexual revolution of the next decade than I had been for my teen years. Within a couple more years, my grandfather's failing health and the need to sell his property to finance his last years cut me off forever from what I viewed as my personal, ancestral homeland. An amorphous, frightening world of multiple unknowns loomed ahead. I returned there in the 1960s for a couple of Levin Cousins Club picnics and my grandfather's funeral in 1966. My father was buried in 1970, my mother in 1979. The unveiling of their headstones within the years of their respective deaths necessitated additional trips to the Alliance Cemetery where they are buried. In the 1990s returning from the New Jersey shore to their home in Maryland, my brother and sister-in-law and I made a side trip there to visit our family's graves, drive past Pop's place, and see the river. Until my pilgrimage in August of 2018, I did not return. By writing this memoir, I am actively engaging in laying most of my ghosts to rest.

Back to the Land

Chapter 8

Looking Back

Two nights after setting the clocks ahead for Daylight Savings, ten days before the Spring Equinox, and two months before my 78th birthday, our 15' by 8', attached solar greenhouse is filling with small kohl crop plants (broccoli, cabbage, cauliflower and kohlrabi), seedlings of lettuce and other greens, peppers and eggplants—all ready to move to the same-size, but free-standing hoophouse, where greens have grown all winter, or to be planted outside. My desire to be fully engaged in this pursuit intensified at the end of January when I could feel a change in the earth's energy. This morning I did not walk as is my wont, but instead engaged in housework. In the afternoon, I transplanted seedlings in the greenhouse and walked thirty minutes before dark, shifting all my energy patterns with the season. I walked out to the county road, a three mile stretch of road that Google Maps cannot locate, and I heard the chorus of high-pitched mating calls of the Spring peepers (*Pseudacris crucifer*) from the "borrow pit" pond at the edge of the road. The county had dug fill and gravel from there some time before we moved here, possibly fifty years ago. For years it functioned as a real pond and habitat to frogs, turtles, snakes and enough fish to feed a great blue heron that we would often see taking flight from the shallow basin. Until about eight years ago when the "pond" became choked with an invasive

185

grass spread either by the neighbor's cattle (the borrow pit is bordered by a strip of our neighbor's land and fenced by him), or perhaps birds; a pair of breeding wood ducks (*Aix sponsa*) stopped there every spring and raised a clutch. The peepers and ducks were signs of hope renewed, and while the ducks have left because there is inadequate water, the peepers' serenade remains, firing up the chorus of male voices every warm day throughout the winter and gladdening the heart of gardeners and others who long for spring.

At this moment in time when our precious blue dot of a planet is so imperiled by global warming/climate change, severe over-population, and pollution of all its natural systems, by geo-political trends of autocracy and outmoded nationalism, anti-Semitism, anti-immigrant nativism and racism, by threats of nuclear annihilation, and proxy wars effecting the most vulnerable populations with famine and total destruction of their lives and cultures, it is easy to despair and difficult to marshal the delight and optimism of my childhood when simple objects and experiences brought me so much pleasure.

It is an easy choice for me to stay at home and immerse myself in the satisfaction and enchantment of my garden, where I cannot completely ignore the world's problems but where they are mitigated by the health of the soil we have built and the bounty it gives back to us for the care we have spent. It is easy to think that I am doing good in the world by tending a garden rather than going places, burning fossil fuels, and spending money to help keep an unsustainable system of runaway consumerism operating.

This past weekend we had a brief but rich visit from a smart, talented, loving, young newly-wed couple who are an important part of our lives. That we have this close relationship is a gift that fills me with hope, optimism, and love because of their capacity for those qualities and the joy they receive from participating in our lives. It is my relationships with loving young people that heal the hurts that are part of all

186

lives, hurts that existed at the time of my early life along with the wonderful moments of innocence, joy, love and belonging. These young daughters and sons of my heart are healers and make me believe in life beyond survival and in viable solutions to the problems that my generation proliferated. As I look back, I see how, despite my childhood and adult awareness that it was not a perfect life, it was largely a life of safety, kindness, gentleness, love and innocence. Now with the awareness of so much abuse, neglect and cruelty in the world, in this day of the Me-Too movement and demand for abusers to be held accountable, I realize how safe and relatively easy, though far from perfect, my childhood was.

Nevertheless, there is one small worm that infects the apple of memory. My mother revealed to me in the last days of my grandfather's life that he was a strict father and that his controlling, demanding ways were the reason that she quit high school in eleventh grade. She wanted independence and freedom, a paycheck of her own earned as salesgirl in a women's clothing shop owned by a mentoring, childless couple, and encouragement to exercise her fashion sense and skill as a talented seamstress and designer. She never told me that he was physically or sexually abusive and never would have permitted me to be alone in the house with him in my childhood or teen years had he demonstrated abusiveness towards her or her sisters.

However, one upsetting and bewildering moment in my relationship with my beloved grandfather exists. It happened one weekend when I was eighteen and getting ready to meet my cousin and friends for an evening activity. Pop had awakened from his nap and stumbled out of his bedroom at the same time I was going to the bathroom. He was literally staggering and muttering, incoherent and physically unstable, and we bumped into each other. He grabbed me and kissed me on the lips, and I pushed away, exclaiming "Pop, Pop, what are you doing?" He didn't seem to realize who I was or who he

187

was, even where we were. Although I was frightened, I was more confused than anything. He had never done anything like that before and never repeated the episode. I gave him as wide a berth as possible, and being the last summer that I spent significant time there, that was easy to accomplish. I never told my mother. Years later after my grandfather and parents were long deceased, I told my brother Mickey. He was, of course, as confused and upset as I had been.

From the moment that this happened, though, I was convinced that it was an anomaly, and that although I could not come up with a rational explanation for his behavior, I knew something was wrong. I've never looked for excuses but craved understanding. At some point much later in life, it occurred to me that his health was already failing, and that he suffered from frequent bladder and kidney infections. What I have learned since then about such persistent urinary tract infections is that they are often accompanied by high fevers and delirium. It is the most plausible explanation for my grandfather's erratic moment, a moment that did not scar me for life but has left me wary of the worm in the apple of pure, joyous memory. I don't want to gloss it over with euphemism nor do I want to exploit it for salaciousness. It happened and I want to be honest and whole in my remembrances while acknowledging that it did not constitute more than a blip on the radar screen. It does not shape my childhood memories and has nothing to do with what I have become. It is only the strangest and single uncomfortable memory of my grandfather that can now be put to rest.

One other thing remains that will not easily be put to rest. Early in the morning of August 12, 2018, I walked from the home of William Levin (great grandson of William and Lena), the house his grandparents I. Harry and Dorothy built in the 1960s, where I had visited in the past, to the cemetery just down the road. I wandered among the tree-shaded older graves, coffee mug in hand, searching for my ancestors. I found the gravestone of great, great grandmother Leah and her second husband and those of my great grandparents, Israel Hersh and Esther Levin, numerous great uncles and aunts, and my mother's cousins. In that cool, peaceful hour as I wandered alone, I could not find the graves of my grandparents and parents. Later in the day of reunion and celebration, with multiple, painful yellow-jacket stings on my bare legs that I had sustained while crossing the lawn and running into a ground nest during the picnic, I searched barefooted in the hot sun and finally found my parents' graves. I walked up and down the pathways between the rows of memorial stones but never found John and Bessie Levin's headstones to place small pebbles upon, signifying enduring memory of the departed. Henry Bermann, the President of Chevra Kadisha (the burial society), had found the location of my parents' stones on his maps and directed me to them where I performed the lithic ritual, but though he searched diligently on his huge paper maps, he could not find those of John H. and Bessie Barish Levin.

Seven months later, although he found numerous Levin graves, including the eldest Levin brother, my young, second cousin was not able to locate my grandparents' stones after a casual search. He also found no trace in the Alliance Cemetery's online locator map. His promise to search more thoroughly and his question to me whether they might be buried elsewhere do not fill me with hope. I remember the day we buried my grandfather in the Alliance Cemetery early in July of 1966. There had been an early dry period, but a gentle, thorough rain had fallen, breaking the early summer drought. John Levin hated drought, and had he been buried in dry, burnt-smelling soil, it would have been a bitter day. When my parents were buried and within a year of the first anniversaries of their deaths, when their headstones were put in place and "unveiled," we visited my grandparents' graves. My niece, at my request, has communicated with the cousin who wishes to have no contact with me. He thought that there were two Alliance cemeteries, but I am unaware of such a possibility. He confessed to not knowing or not remembering where they were buried. Then in April of 2019, my cousin David Levin also searched, both the cemetery and the online locator, with the same results. That my grandparents seem to have vanished from the cemetery is unthinkable; it plucks me out of my place in the real world and drops me into a twilight zone, an alternate universe of nightmarish twists, turns and final dead ends. Sometimes at night when sleep won't come and the past tugs at my pajama sleeve, fearful imaginings plague me. I wonder if my grandfather had committed some egregious action that was swept under the rug, something taboo enough to have banished him and my grandmother from the tribal cemetery. Or had my cousin surreptitiously removed their coffins to another cemetery for an unknown and inexplicable reason? I will never know and have no options left to me to locate them. I am left grieving for their loss.

But I am alive and have richly relived my childhood and teen memories as I have written this memoir. I also live intensely in the now, a now ruled by natural seasons, cycles and rhythms. As I endeavor to neatly tie up my memories of long-gone times and reflect about their meanings and how those years shaped my life, my perception of myself as a metaphoric sport of nature hearkens back to the myths and realities of the Alliance Colony. One of the reasons that I never did well with careers is that I am a natural anarchist. I quickly tire of artificially imposed schedules and am more immune to illness when I do not fight the urge of my body and mind to go with natural rhythms: the seasons, the weather, the phases of the moon. However energetic or tired my body is; whatever muse pops into my consciousness and takes hold of me, I try to do its bidding to find the right outlet for my creative energy. When allowed to live to an adequate degree by this natural spontaneity, I am more able to accommodate others and find a suitable, cooperative pattern for living and working together; I am physically and mentally healthier. To me this seems to be an inherent, natural way to live, but does not adapt well to the world of capitalism and consumerism. To choose this life path means that progress and wealth are never more important than self-awareness and fulfillment. By this, I do not mean selfishness or egotism but an embrace of a holistic life.

For my husband and me, life over the last forty-five years has revolved around a focal point of existence: growing food spring through fall and making things through the winter. I am, like my mother, a seamstress and clothing designer. I am also a quilter and handweaver, a maker of beautiful and functional things by hand, and a sometimes painter. I am not a farmer and have never needed to grow cash crops, but my husband and I grow and preserve enough food to slash our grocery costs significantly. We eat well—delicious, healthful, mostly organic food, as local as possible, mostly a vegetarian diet with flexibility and occasional exotic treats. We would be eating above our economic means but for the volume of the food we grow. In the growing and harvesting seasons, we work many hard hours in the garden and have little time for anything else. We usually cannot take vacations during gardening season or go away for more than an overnight. The parameters of a farming life shape our lives. The winter is the time when all the creative juices that fermented over the growing season get put into physical, material form. The structure of the lives of my ancestors and their fellow settlers, which meant life or death for them, is the armature of my life to sculpt a creative, living form. I have the economic freedom to do that even though I do not have wealth. It is a choice I have made to become a self-realized human being, and I am grateful for the freedom while fully accepting of the limitations.

Among the Jewish farmers of the Alliance Colony in 1899, one J. C. Reis wrote to *The Jewish Exponent* of Philadelphia in answer to the editorial question posed in an article "Can

Jews Be Farmers?" His answer is a resounding "Yes," and his description of several of the more successful farmers is laudatory. One of his "good farmers" was my great grandfather Hersh Levin who owned a 41-acre "good farm in productive state." In 1899 there were ten people in Hersh's family, and he and Grandmother Esther had been there for eighteen years, whole-heartedly preferring farming to city life. Mr. Reis saw my grandfather and great uncles as "industrious and promising young farmers" and reported that Hersh Levin acknowledged that all his sons "like farming. If I only could give them each a farm, but my farm is not large enough to parcel up in this way." He goes on to say that he had recently paid for the building of a new barn and the wedding of one of his daughters—a drain upon his resources but one which he would alleviate with continued hard work.[1]

Upon first reading this interview with my ancestor—close to the end of writing this memoir—I felt the hairs stand up on the back of my neck. My great grandfather, about whom I had never felt such a strong sense, seemed alive and present to me. I wanted to ask him, "Well, Zeyde, what do you think of your son John's only granddaughter who, 120 years after J. C. Reis interviewed you, has been a kind of subsistence farmer for nearly half a century? How would you have felt about leaving some acreage to her? And your son William's great grandson—his namesake—and young Will's wife trying to bring back Jewish farming to this community where you settled—what do you think of them?"

Through these questions alone, I deem the Alliance Colony a great success, complete with my ancestors' part in it, and their legacy to me and my cousins, especially my second cousin once removed, William. While I, myself, am not an observant Jew, I view as a good thing William and Malya Levin's efforts to create a renaissance for Jewish farming in the area and once again bring a significant Jewish presence to Pittsgrove Township of Salem County, New Jersey. I am just as excited about the rural and urban renewal of black farming as a reclamation of liberation through the soil. Leah Penniman, the author of *Farming While Black*, has worked for twenty-five years at Soul Fire Farm in Grafton, New York, as an organic farmer. She is also a teacher and an activist supporting racial justice in the food system. Jewish farmers of today employ kosher methods of slaughter on small-scale sustainable farms. Many cultivate heirloom grains and foods that were grown in ancient Israel as well as contemporary market favorites. I assume that black farmers follow similar paths, growing foods that came from Africa and have been incorporated into American cuisine via Southern foodways.

Until about 1850, most people in the United States lived on or had close connections to farms. People grew and prepared their own food. In communities with strong ethnic identities, the foods they grew and ate had distinctive characteristics. To this day heirloom tomatoes, for example, bear colorful, intriguing names that identify the countries from which they traveled with emigrant populations as saved seeds to take root in the new world in farms and backyard gardens: Amish Paste, Black Prince and Black Krim (both from Russia), German Johnson, Japanese Trifele, Nepal, and Polish Linguisa, to name a few. Industrial scale farming, urbanization and fast food have in the past homogenized many immigrant foodways, which had comprised American cuisine, until recent movements began to pump new life into them. Look at any seed catalogue, stroll the streets of any mid- to large-sized American city, and you will find ethnic food trucks and restaurants. Even small-town grocery stores carry specialty and available ethnic vegetables and fruits. While there is the danger of cultural appropriation of simple ethnic foods by rock-star chefs and high-priced restaurants, people are finding their roots through food as well as other channels.

All people deserve a connection to the land, even if "the land" is more theoretical in nature and their food is grown vertically in inner city buildings. Will Allen, of the now defunct Growing Power Farm Projects, is a black urban farmer who preceded Leah Penniman. A MacArthur Foundation Genius Grant recipient in 2008, as well as a Ford Foundation grant recipient, Allen had played college and professional basketball before returning to his farming roots and establishing several sites for Growing Power's vertical, urban farms. Growing Power became bankrupt, not because its model is a failure, but because the predominant, subsidized model of non-sustainable, corporate farming maintains an iron grip on food production while failing human life and culture, as well as the environment. Vertical, urban farming produces high quality, sustainable, clean food. Today more and more young people are seeking lives as urban and rural farmers on a small, environmentally responsible scale. Allen, Penniman, my cousin Will and his friends in Alliance, New Jersey, and young urban and rural farmers in Arkansas and elsewhere are my heroes. I am happy to have lived a life that has been compatible with their practices. Food sovereignty and sustainable agriculture are blending in many new ways, developing paradigms on a more human scale, carrying hope for the future.

I owe my back-to-the-land life of the past forty-three years to my life partner and husband and to my brave Russian Jewish ancestors who made the difficult journey to South Jersey for their back-to-the-land life of freedom in America. Without my grandfather John H. Levin's embodiment of that lifestyle, I would never have had the prototype of being a "free farmer on my own soil." For all of this, my gratitude overflows.

Epilogue

December 19, 2019, was a mild and mostly sunny day in Arkansas. The occasional cloud made the drive to Harrison easier on my eyes than a bright, fully sunny day and less stressful than a rainy day would have been. I was gone all day, having stopped for my regular eight-month haircut in Marshall before continuing on to Harrison. Harrison, Arkansas, has an unsavory reputation as a white supremacist town, but is fortunate to have many residents who fight to change that ugly past through a more accepting, inclusive and progressive social and political outlook and through activism. We are happy to count many of these open-minded people as our friends.

Harrison is also the nearest town large and diversified enough to be a destination for multi-purpose shopping trips: I can go to a well-stocked health food store, take my choice of liquor stores in a county that until four or five years ago was as dry as the county we live in, avail myself of several good grocery stores, and purchase printer ink cartridges and other hard to find products at Walmart. At fifty miles distant, it's close and small enough to be a relatively easy journey through the beautiful scenery of Ozark hills and hollers, if one can time the trip to avoid rush hour traffic. On my way home, I stopped in Marshall again to visit a former neighbor who had recently had a serious stroke and, against all odds, is starting to make a significant comeback. This tough old woman is not someone I consider a close friend because of many major differences between us, but she is a neighbor of value, and around here, neighbors take care of neighbors.

After I got home and unloaded the car, fed the yowling cats, and walked for thirty minutes—before cooking myself a nice meal and pouring a glass of wine—I booted up my laptop and opened g-mail to find a life-outlook changing email. In South Jersey the weather was not warm and pleasant. Tom Kinsella, editor extraordinaire and, now, family friend forever, had driven to the Alliance Cemetery because he was in the general area for a meeting. Soon, upon entering the cemetery on a cold, windy day, Tom, who is a professor of literature, director of the South Jersey Culture & History Center, and the Samuel and Elizabeth Levin Director of the Alliance Heritage Center at Stockton University, found several Levin graves, the same ones I saw in August of 2018: my great grandparents, my great, great grandmother, great uncles and aunts, and multi-generational cousins. My parents and my mother's sisters are buried in another part of the cemetery, their married names not Levin. My brother's remains were cremated and his ashes scattered elsewhere. Missing from my visits to the cemetery in the summer of 2018 and from the Alliance Cemetery maps and on-line locator were the graves of my maternal grandparents, John Levin and Bessie Barish Levin, a lack that I could not quite put to rest. Two of my second cousins searched for me as well, to no avail.

That Tom looked and found these simple foot stones when they seemed certainly to be mysteriously and horribly gone and, in the Jewish tradition, placed a small, ordinary pebble upon each of the gravestones, signifying that the living remembered the dead, touched me and my husband deeply. That he understands and has such deep respect and affection for the vast collection of material from the Jewish-immigrant agricultural settlements of southern New Jersey is a testimony to his skill as an academic and archivist, but his small kindness speaks so eloquently of what kind of man he is: a

196

mensch par excellence. In the obligatory formulae of the dark Eastern European fairytales that my ancestors may have told to misbehaving children—if I were a young nubile woman, I would owe my first-born child to Tom for performing a deed that no one else was able to do. Fortunately, my ancestors emigrated from the Russian-occupied Ukrainian Pale in 1882, leaving behind tyrannical obligation to the laws of the Tsar and gaining their freedom. And what seemed to me to be an impossible happy ending, in ever finding my grandparents' graves, has come to pass, thanks to Tom. My story is now as complete as it will ever be, and I am content to close the cover on the final chapter of this book.

Back to the Land

My Family

Back to the Land

(Left to right) Baba Esther Levin, Ruth's great grandmother; Hannie Coltun, Ruth's great aunt (standing); Baba Leah Levinson, Ruth's great, great grandmother; Helen Coltun (sitting). Courtesy of Rich Brotman.

The Levin brothers in the early 1900s. (Standing, left to right) John H. Levin, Abe; (seated, left to right) Manny, William, Samuel. Courtesy of Marsha Levin Schumer.

The Levin brothers in the early 1900s. John H. Levin is far right. Courtesy of Marsha Levin Schumer.

John H. Levin "Pop" standing in front of his bungalow on Gershal Avenue, Norma, c. 1956–1957.

(Top) Sylvia Levin Weinstein, June 1940. (Bottom) Meyer "Marty" Weinstein preparing smoked fish platter for the Levin cousins' club meeting in the Weinstein Philadelphia row house, May 1955.

Sylvia Levin Weinstein and Ruth at Norma Beach, August 1943.

(Top) Family and friends in tube, August 1943. Left to right, unknown female cropped, Bruce Bailine, Aunt Jean Bailine holding two-year-old Ruth, Mickey; in front, Toby Berkowitz, a family friend. (Bottom) Mickey and Ruth with puppies at Pop's house, c. 1946.

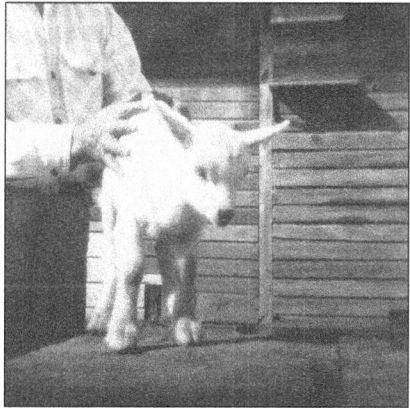

(Top left) John H. Levin "Pop" with neighbor Laura Lehman and Ruth, c. 1946. (Top right) Cousin Bruce Bailine, Mickey Weinstein and Ruth, c. 1946. (Bottom left) Mickey and Ruth, c. 1947. (Bottom right) John H. Levin with goat kid.

(Top) Levin cousins' club meeting, 1954. Courtesy of Marsha Levin Schumer.
(Bottom) Fourteen-year-old Ruth hanging from tree branch at Pop's, June 1955.

Ruth Weinstein at Norma Beach, 1956.

(Top) Joe McShane plowing the garden with the donkey. (Bottom) The Arkansas garden c. 1981.

(Top) Ruth with goats, c. 1981. (Bottom) Ruth Weinstein and Joe McShane, 2020.

Appendix: Original Settlers

Eli and Ethel Abramovitz
Eli and Feigeh Bakerman
Moses and Ethel Bayuk
Abraham and Channah Leah Berman
William and Becky Cohen
Hersh and Jennie Coltun
Joseph and Rachel Diamond
Jacob and Rebecca Ecoff
Chaim and Bessie Goldman
Nissan and Molka Greenspan
Abraham and Duba Grutsky
Simcha and Sarah Helig
Joseph and Yenta Kleinfield
William and Lizzie Kolman
Zurach and Esther Konowitz
Isaac and Golda Krassenstein
Labe and Bayla Kuden
Hersh and Rivka Kutzibow
Israel Hersh and Esther Levin*
Leapa and Toba Levinsky
Berel and Leah Levinson*
Labe and Toba Riva Levinson
Henry and Rose Levy
Sholom and Pearl Luberoff
Simcha and Pearl Luborsky
Chaim and Sarah Mennies

Labe and Rachel Moyd
Israel and Feigeh Opachinsky
Lazar and Mindel Perskie
Jacob and Golda Rosenberg
Yonah and Anna Rosenfeld
Jacob and Anna Rosinsky
Joseph and Feigeh Rothman
Joseph and Deborah Rudnick
Solomon and Frima Salonsky
Moshe and Ruchel Serebrenick
Hersh and Rose Silberman
Chaim Hersh and Sima Liba Silberman
Lazar and Bessie Staver
Eli and Rita Gitel Stavitsky
Moses and Bayla Strasnik
Pesi and Brucha Tolchinsky
Naphtula and Deborah Yosep
Joseph and Rose Zager

* Levin family ancestors

Acknowledgments

First thanks go to Jay Greenblatt, attorney, historian, visionary and fellow descendant of the Alliance Colony, for his enduring commitment to the Alliance Colony Foundation and humanity, for facilitating the 136th anniversary celebration, something I wished for so ardently, and for cheerfully answering my barrage of questions. To the Jewish Federation of Cumberland, Gloucester and Salem Counties of New Jersey for their support of the Colony Foundation and the reunion picnics. Professor Tom Kinsella and his students at Stockton University for archiving the records of the colony; to him and the university, many are grateful.

To others whose work is part of the tapestry of the Alliance Colony, many thanks. Those I have been fortunate to meet are Rich Brotman and Susan Kehnemui Donnelly, documentary filmmakers; Jarrett Ross, Jewish genealogist; Howard Jaffe, farmer, historian and preservationist.

To my first readers who inspire me to ruthless self-editing of my glittering and clouded memories: Donna Cave, thorough and thoughtful reader, excellent editor, and long-time supporter of my writing life; Janie Pritchett-Clark who knows the longing for place and grooves on shared food, wine and writing; Toni Newby whose intellect and courageous compassion go beyond clear and loving editorial advice; to the memory of Karen Hayes, host of Second Thursday Poetry Night at Guillermo's in Little Rock and beloved poet, muse, friend of many, for luring me back to a great open mic; and to Poetluck potluck regulars and residents and former director Linda Caldwell, at Writers' Colony at Dairy Hollow in Eureka Springs, Arkansas—a second home. Linda's successors, Michelle Hannon and Chad Gurley and the board of directors at WCDH carry on the good work. Specific thanks to Woody Barlow, a friend from the colony, for reading and commenting on this work.

To the Levin cousins who have encouraged this endeavor and/or have played a big part in my past: specifically Liz Kelner Pozen, old friend and travel partner, fellow poet, painter, and storehouse of family lore and history; her first cousin, Bob Levin, retired attorney, writer and resident of Berkeley, for his knowledge and advice; to Marsha Levin Schumer, a talented artist, for her generosity with family photographs and excellent memory for the family history; to all the rest, including those who are now gone.

To my family of choice: friend Laura Daly, my "daughter of the heart," with whom sharing tea and conversation always inspires me to greater understanding and compassion.

And finally, gratitude and love to my husband, Joe McShane, without whom this life would not have been possible, as necessary to me as air.

To Professor Thomas Kinsella and the excellent staff at Stockton University and the Alliance Heritage Center—research fellows Sara Brown and Ray Dudo and graphic artist Jena Brignola (who designed the cover)—a huge thank you for publishing this book and doing such a great job with editing, formatting, design and everything else that I could not accomplish on my own. Tom's expertise has always come wrapped in his friendly, supportive kindness.

Any errors in the history of the Russian Jews and their migration or of the Alliance Colony are mine. Some differences of fact and interpretation exist among the descendants of the Levin Family. I am reporting what various Levin cousins have shared with me without preference for an absolute Truth in the family history. I am sharing my own memories, which are altogether imperfect, like a beautiful, but badly moth-eaten paisley shawl. I have not fabricated anything for this account, but if my version does not the match time, place and events remembered by others, I am sorry. This is neither a work of fiction nor a lab report, but a work of creative non-fiction over which I have labored long and hard. The mistakes are part of my story and not an intent to mislead. And, if I have neglected to thank anyone who contributed to the creation of this book, I offer my deepest apologies.

Back to the Land

Notes

Chapter 1
Why on Earth Write a Memoir?
1. *YOVAL: A Commemoration of the First 125 Years of the Jewish Farming Colonies of Alliance, Norma & Brotmanville, New Jersey; August 26, 2007* (The Reunion Committee, 2007), 1.

Chapter 2
History of the Alliance Colony
1. "Pogroms in the Russian Empire," *Wikipedia*, en.wikipedia. org/wiki/Pogroms_in_the_Russian_Empire.
2. Ellen Eisenberg, *Jewish Agricultural Colonies in New Jersey, 1882–1920* (Syracuse, NY: Syracuse University Press, 1995), 6–7.
3. Richard Brotman, *First Chapter in a New Book: The Story of Brotmanville and the Alliance Colonies*, DVD documentary (New York: Richard Brotman, 1982 and 2007), interview with Lillian Greenblatt Braun.
4. Ibid., interview with I. Harry Levin.
5. Joseph Brandes, *Immigrants to Freedom: Jewish Communities in Rural New Jersey since 1882* (Philadelphia: University of Pennsylvania Press, 1971), 17.
6. Leslie F. Larson Bennett, *Zion in the New World: The Lubarksys Find The "Goldene Medina,"* kehilalinks.jewishgen. org/NJ_Farms/Lubarsky_Alliance.pdf.
7. "Am Olam," *Encyclopedia.com*, www.encyclopedia.com/ religion/encyclopedias-almanacs-transcripts-and-maps/am-olam.

8. Ibid.
9. Ibid.
10. "New Odessa Colony," *The Oregon History Project*, oregonhistoryproject.org/articles/historical-records/new-odessa-colony/#.XXL1_JNKjEY.
11. Uri D. Herscher, *Jewish Agricultural Utopias in America, 1880–1910* (Detroit: Wayne State University Press, 1981), 52–54.
12. Eisenberg, *Jewish Agricultural Colonies in New Jersey*, 52.
13. Ibid., 53.
14. Brandes, *Immigrants to Freedom*, 162.
15. Eisenberg, *Jewish Agricultural Colonies in New Jersey*, 53.
16. Ibid.
17. Ibid.

Chapter 3
The Alliance Colony Benefactors
1. Eisenberg, *Jewish Agricultural Colonies in New Jersey*, 92–93.
2. "Jewish/Israel Organizations: Alliance Israelite Universelle," *Jewish Virtual Library*, www.jewishvirtuallibrary.org/alliance-israelite-universelle.
3. Tom Kinsella, *Heilprin's Appeal to the Jews* (Galloway, NJ: Alliance Heritage Center, 2019). This brief treatis is available in the Alliance Heritage Center archive, Special Collections & Archives, Bjork Library, Stockton University.
4. "Hebrew Emigrant Aid Society," *Wikipedia*, en.m. wikipedia.org/wiki/Hebrew_Emigrant_Aid_Society.
5. "Baron Maurice de Hirsch," *Jewish Virtual Library*, jewishvirtuallibrary.org.
6. "BILU," *Jewish Virtual Library*, jewishvirtuallibrary.org.
7. "Jewish Colonization Association (ICA)," *Jewish Virtual Library*, jewishvirtuallibrary.org.

Chapter 4
Life in Alliance
1. Brotman, *First Chapter in a New Book.*
2. Eisenberg, *Jewish Agricultural Colonies in New Jersey*, 115.
3. Herscher, *Jewish Agricultural Utopias in America*, 74.
4. "The Jewish Farmer: Will they Make Farmers," *The Sun*, www.museumoffamilyhistory.com/lia-hww-farming-sun.htm.
5. Eisenberg, *Jewish Agricultural Colonies in New Jersey*, 102–103.
6. "Jewish Agricultural (and Industrial Aid) Society," www.jewishvirtuallibrary.org/jewish-agricultural-and-industrial-aid-society.
7. "The Jewish Farmer: Will they Make Farmers," *The Sun*, www.museumoffamilyhistory.com/lia-hww-farming-sun.htm.
8. Ibid.
9. Herscher, *Jewish Agricultural Utopias in America*, 106.
10. Ibid., 105.
11. Brandes, *Immigrants to Freedom*, 145.
12. Ibid., 152.
13. Ibid., 146.
14. Ibid., 66.
15. Brotman, *First Chapter in a New Book.*
16. Eisenberg, *Jewish Agricultural Colonies in New Jersey*, 166.
17. Sidney Bailey, "The First Fifty Years," *YOVAL: A Symposium Upon the First Fifty Years of The Jewish Farming Colonies of Alliance, Norma and Brotmanville, New Jersey* (August 1932; republished 1982), 16.
18. Todd R. Sciore, "Unlikely Farmers: Tokens of the Allivine Canning Company," *The Numismatist* (January 2018), 58–59; find a slightly revised version in *SoJourn: A Journal Devoted to the History, Culture and Geography of South Jersey*, 3.2 (Winter 2018/19), 55–62.
19. Brandes, *Immigrants to Freedom*, 77.
20. Herman Eisenberg, "The Golden Age," *YOVAL* (1932), 25.
21. Ibid., 26–27.

22. Eisenberg, *Jewish Agricultural Colonies in New Jersey*, 139.
23. Helig in Brotman, *First Chapter in a New Book*.
24. *YOVAL* (1932).
25. Bailey, "The First Fifty Years," *YOVAL* (1932), 18–19.

Chapter 5
The Levin/Levinson Family History
1. Elizabeth Rudnick Levin, "Pioneer Women of the Colonies,"
YOVAL (1932), 31.
2. Ibid., 31.

Chapter 6
The Home of John H. Levin, "My Redneck Zeyde"
1. "Vinelands's Legendary Palace," *The Daily Journal*,
December 4, 2015, www.thedailyjournal.com.
2. Julia Klein, "The Long Journey into the Light on Jersey
Farmlands," *Philadelphia Inquirer*, August 12, 1979, 156,
www.newspaper.com.

Chapter 7
The Maurice River, 1943 to 1959
1. "Maurice River," *Wikipedia*, en.wikipedia.org/wiki/
Maurice_River.
2. "Welcome to the Maurice River," *Maurice River*, maurice-
river.igc.org/.
3. Hellen A. Oxman, "Memories of Pleasure in Summer,"
Philadelphia Inquirer, August 15, 1982, 11.
4. Elizabeth Rudnick Levin, "Pioneer Women of the Colonies,"
YOVAL (1932), 31.

Chapter 8
Looking Back
1. Moses Klein, *Migdal Zolphim & Farming in the Jewish
Colonies of South Jersey* (Galloway, NJ: South Jersey Culture
& History Center, 2019), 201.

Bibliography and Recommended Reading

Bayuk Rappoport Purmell, Bluma and Felice Lewis Rovner. *A Farmer's Daughter: Bluma* (Los Angeles, CA: Hayvenhurst Publishers, 1981).

Brandes, Joseph. *Immigrants to Freedom: Jewish Communities in Rural New Jersey since 1882* (Philadelphia: University of Pennsylvania Press, 1971).

Clark, Eugenia. "A Farmer's Daughter Recounts the Early Days of Life in the Alliance Colony," *Philadelphia Inquirer*, August 15, 1982. 10–12.

Clark, Eugenia. "At 100, Alliance Celebrates Legacy of Its Forebears," *Philadelphia Inquirer*, August 15, 1982. 10, 12.

Eisenberg, Ellen. *Jewish Agricultural Colonies in New Jersey: 1882–1920* (Syracuse, NY: Syracuse University Press, 1995).

Goldstein, Philip Reuben. *Social Aspects of the Jewish Colonies of South Jersey* (Philadelphia: University of Pennsylvania Ph.D. thesis, 1921).

Herscher, Uri D. *Jewish Agricultural Utopias in America, 1800-1910* (Detroit: Wayne State University Press, 1981).

Klein, Moses, et al. *Migdal Zophim & Farming in the Jewish Colonies of South Jersey* (Galloway, NJ: South Jersey Culture & History Center, 2019).

Levin, Bob. *CHEESESTEAK: The West Philadelphia Years, (a rememboir)* (Berkeley: Spruce Hill Press, 2018).

Oxman, Helen A. "Memories of Pleasure in Summer," *Philadelphia Inquirer*, August 15, 1982, 11.

Pozen, Liz Kelner. *The Heart of The Family: Poems and Paintings* (Hollis: Somerset Press, 2016).

YOVAL: A Symposium upon the First Fifty Years of the First Jewish Farming Colonies of Alliance, Norma, and Brotmanville (New Jersey. Published privately for the 50th anniversary celebration, 1932; republished 1982).

YOVAL: Alliance Colony 1882–2007, A Commemoration of the First 125 YEARS of the Jewish Farming Colonies of Alliance, Norma & Brotmanville, New Jersey (The Reunion Committee, 2007).

About the Author

In a long ago former life, Ruth Weinstein taught high school English in Philadelphia, English as a Foreign Language in Japan, and English as a Second Language in Arkansas. For nearly forty-five years, however, she has focused on organic gardening; writing poetry, essays and memoir; and making both functional and art pieces in a variety of textile media. She has won awards for poems about how gardening and food affect personal and community relationships. She and her husband live on their forty-acre wooded homestead in the Arkansas Ozarks.

Chapter text set in 12-point Sabon LT Pro;
poetry and table of contents in 12-point Cronos Pro.
South Jersey Culture & History Center Regional Press
for the Alliance Heritage Center at Stockton University.

www.ingramcontent.com/pod-product-compliance
Lightning Source LLC
Chambersburg PA
CBHW021051090426
42738CB00006B/293